Mechanical Appliances, Mechanical Movements and Novelties of Construction

Gardner D. Hiscox

Dover Publications, Inc.
Mineola, New York

Note

This book is the second in a two volume series on mechanical design by Gardner D. Hiscox. The first, *1800 Mechanical Movements, Devices and Appliances,* is also available from Dover Publications. Go to *www.doverpublications.com* for more information.

Bibliographical Note

This Dover edition, first published in 2008, is an unabridged republication of the work originally published in 1916 by Munn & Company, Inc., New York.

Library of Congress Cataloging-in-Publication Data

Hiscox, Gardner Dexter, 1822?–1908.
    Mechanical appliances, mechanical movements, and novelties of construc-
tion / Gardner D. Hiscox.
        p. cm.
    Reprint. Originally published: New York : Munn & Co., 1916.
    ISBN-13: 978-0-486-46886-0
    ISBN-10: 0-486-46886-0
    1. Mechanical movements. I. Title.

TJ181.H7 2008
621—dc22

2008019640

Manufactured in the United States of America
Dover Publications, Inc., 31 East 2nd Street, Mineola, N.Y. 11501

Preface.

THE many editions through which the first volume of "Mechanical Movements" has passed is more than a sufficient encouragement to warrant the publication of a second volume, more special in scope than the first, inasmuch as it deals with the peculiar requirements of various arts and manufactures, and more detailed in its explanations, because of the greater complexity of the machinery selected for illustration. Despite the greater simplicity of the devices which have been pictured and briefly explained in the first volume, the appliances described in this second volume can be just as easily understood, the text having been so worded that no insuperable difficulties are presented to the reader of average mechanical knowledge. More extensive though it may be than "Mechanical Movements," the present work by no means exhausts the subject. Many an apparatus has been omitted, either because limitations of space have intervened, or because of the impossibility of securing adequate details of construction. The machines incorporated, however, cover so vast a mechanical field and have been so carefully selected to supply the needs of the student seeking general information, that they will be found fairly representative of the power devices used in old and modern industries. Mechanical intelligence may well be deemed to have found its highest expression in the contrivances that are illustrated and described in these pages.

Although the author has not the slightest desire to encourage the hopeless pursuit of perpetual motion, he has, nevertheless, thought it advisable to dwell at some length on the exceedingly ingenious means devised by misguided inventors in their endeavors to solve an unsolvable problem. The pages in which perpetual motion machines are described may induce those who still believe in reaching this *ignis fatuus* to bend their energies in causes more worthy of their zeal. Moreover, it may be that some of the mechanical movements which have been evolved by the perpetual motion inventor, although they may not attain the end sought by him, may still be applied with profit to his instruction in true mechanical principles and to avoid the errors committed in the search on the lines of this folly of past centuries. This in itself is a sufficient justification of the insertion in this volume of the section on perpetual motion.

The deeper we delve in the research for novelty and variety in the present field of mechanical design, the more we see the possibilities of human ingenuity. The facility and power of construction shown in the complicated mechanism of the past augur well for the future of inventive genius.

GARDNER D. HISCOX.

January, 1914.

CONTENTS.

SECTION I.

MECHANICAL POWER LEVER.

SECTION II.

TRANSMISSION OF POWER.

SECTION III.

MEASUREMENT OF POWER, SPRINGS.

SECTION IV.

GENERATION OF POWER, STEAM.

7

SECTION V.

STEAM POWER APPLIANCES.

SECTION VI.

EXPLOSIVE MOTOR POWER AND APPLIANCES.

SECTION VII.

HYDRAULIC POWER AND APPLIANCES.

SECTION VIII.

AIR-POWER MOTORS AND APPLIANCES.

SECTION IX.

GAS AND AIR-GAS DEVICES, ETC.

SECTION X.

ELECTRIC POWER AND DEVICES.

Electric Cable-Making Machine—Chloride Accumulator—Electric Wire Insulating Device—Electric Wire Doubling Device—Cable Cover, Braiding Machine—Wire-Covering Machine—Shunt-Wound Dynamo—Shunt Dynamos connected in series—Short and Long Shunt—Separately Excited Dynamo—Compound Wound Dynamos—Multi-Speed Electric Motor—Drum Controller—Commutator Construction—Spring Binding Post—Electric Transformer—Recording Ampere Meter—Novel Arc Lamp—Searchlight Mirror—Electric Engine Stop—Series Arc Lighting Circuit—Rotating Electric Furnace—Electric Blowpipe—Electric Furnace—Tandem Worm-Gear Electric Elevator—Electrically Driven Sewing Machine—Electric Motor Worm-Driven Pump—Electric Incubator—Electrical Soldering Copper—Electric Welding Apparatus—Electric Welding—Electric Revolving Crane—Electro-Magnetic Track Brake—Electro-Magnetic Clutch—Wireless Telegraphy—Automatic Trolley-Wheel Guard—Electric Lighting System—Electrically Heated Chafing Dish—Vibrating Electric Bell—Printing Telegraph—Electric Fire-Alarm System—Electric Tree-Felling Machine—Electric Trumpet—Electric Blue Print Machine—Demagnetizing a Watch—Electric Curling-Iron Heater.

SECTION XI.

NAVIGATION, VESSELS, MARINE APPLIANCES, ETC.

Curious Boats—Greenland Kayak—Racing Yachts—Ancient Feathering Paddle Wheel—Types of Propellers—Screw Propeller—Sheet Metal Propeller—Feathering Blade Propeller—Twenty-five-Foot Launch—Bicycle Catamaran—Bicycle Gear for a Boat—The Manipede Catamaran—Types of Shallow-Draught-Screw-Propelled Boats—Dirigible Torpedo—Automobile Torpedo—The Holland Submarine Boat—Reversing Clutch—Ice Boat—Submarine Cable Grapnel—Steam Sounding Machine—The Drag Steering Gear—Rope Hitches—Knots and Splices—Bell Buoy—The Whistling Buoy—Lighting Buoy—Fog Whistle—Fish Way—Floating Breakwater—Nets and Seines.

SECTION XII.

ROAD AND VEHICLE DEVICES, ETC.

Road Grading Wagon—Traction Wheel—Dumping Wagon—Differential Speed Gear—Automobile Steering Gear—Ratchet Brake Lever—Automobile Change Speed Gear—Automobile Steam Engine—Types of Motor Bicycles—Steam Surrey—Steam Freight Wagon—Steam Dray—Interchangeable Automobile.

SECTION XIII.

RAILWAY DEVICES AND APPLIANCES, ETC.

Block and Interlocking Signals—Railway Signals—Trolley-Car Sander—Locomotive Sander—Multiple Plate Friction Clutch—Types of Trolley-Car Trucks—Types of Rack-railway Locomotives—Fare-Recording Register—Cable Grip—Railway Track Brake—Rolling and Compressing Steel Car Wheels—Reversing Car Seat—Four-Spindle Rail Drill—Crank-Pin Turning Machine—Extension Car Step—Trolley Replacer—Car Coupler—Bulldozer Press.

SECTION XIV.

GEARING AND GEAR MOTION, ETC.

Novel Worm Gear—Swash-Plate Gears—Stop-Gear Motion—Volute Tappet Gear—Geared Reversing Motion—Elliptic Linkage—Interrupting Cam-Gear Motion—Circular from Reciprocating Motion—Crank Substitute—Sun and Planet Motion—Intermittent Rotary Motion—Friction Gear—Parallelism from Circular Motion—Circularly Vibrating Motion—Differential Speed Gear—Epicyclic Train—Transmission Gear—Variable Speed Friction Gear—Variable Speed Gear—Driving Gear for a Lathe—Variable Drive Motion.

SECTION XV.

MOTION AND CONTROLLING DEVICES, ETC.

Parallel Motion—Three-Point Straight-Line Linkage—The Dead Center Problem—Crank Substitute—Short-Range Walking Beam—Turning a Square by Circular Motion—Double-Link Universal Joint—Change Speed Pulleys—Multiple-Shaft Driving Device—Reciprocating with Stop Motion—Reciprocating Motion—Reciprocating Into Rotary Motion without Dead Centers—Right-Angle Coupling—Reversible Friction Ratchet—Friction-Plate Clutch—Friction Clutch—Expanding Wrench or Chuck—Multiple Ball Bearings—Shaft-Thrust Ball Bearings—Bicycle Ball Bearing—Ball-Bearing Castor—Spring Motor—Weight Driven Motor—Swing Motor—Ammonia Compressor—Coin-in-the-Slot Gas Meter—Spiral Fluting Lathe—Pantographic Engraving Machine—Geometrical Boring and Routing Chuck—Rose Lathe—Plantariums—The Phenakisto-scope.

SECTION XVI.

HOROLOGICAL TIME DEVICES, ETC.

Electric Pendulum—Electric Clock Controller—Repeating Clock Escapement with Electric Pendulum—Electric Ratchet—Solar and Sidereal Clock—Novel Clock—Electrical Correction of Clocks—Long-Distance Telegraph-Clock Correction—Flying-Pendulum Clock—Self-Winding Synchronizing Clock.

SECTION XVII.

MINING DEVICES AND APPLIANCES.

Mining Lamp—Well-Boring Tools—Prospecting Diamond Drill—Assay Ore Crusher—Ore Roasting Furnace—Magnetic Metal Separator—Magnetic Separator—Quartz Pulverizer—Ore Washing Tower—Automatic Ore Sampler—Pneumatic Concentrator—Ore Car on a Transfer Truck—Dry Placer Gold Separator—Dry Gold Mining Machine—Gold Amalgamator—Sheave Wheels for Gravity Planes—Briqueting Machine—A Briqueting Plant—Coal-Washing Jig—Propeller Pump Agitator—Coal-Handling Plant—Method of Change Direction.

SECTION XVIII.

MILL AND FACTORY APPLIANCES AND TOOLS, ETC.

Machine-Made Chains—Suspending Grip—Universal Dog—Drill Chuck—Brick Clamp—Combination Tools—Easily Made Steam Whistle—Gasoline Heated Soldering Copper—Pulley Balancing Machine—Lubricating Drill—Expanding Drill—Taper Attachment to a Lathe—Taper Turning Attachment—Centering Device for a Drill Press—Boring Elliptic Cylinders—Crane Truck—Centrifugal Separator—Blacksmith Helper—Belt-Driven Forging Hammer—Eye-Bending Machine—Angle Iron Bending Machine—Pipe-Bending Machine—Rolled-Thread-Screw Machine—Power Hack-Saw—Seamless Tube Machine—Metal Band-Saw—Hand-Screw Tire-Setting Machine—Hydraulic Tire-Setting Machine—Automatic Furnace—Gas-Heated Hardening and Tempering Furnace—Tempering Bath—Down-Draught Gas-Melting Furnace—Oil or Gas Fired Forge—Melting Furnace—Duplex Melting Furnace—Open Hearth Steel Furnace—Hot-Metal Mixer—Kerosene-Oil Melting Furnace—Petroleum Forge—Petroleum Melting Furnace—Petroleum Fired Reverberatory Furnace—Plate Hardening Machine—Dovetailing Machine—Diamond Millstone-Dressing Machine—File-Cutting Machine—Dovetails—Mortising Dovetail Machine—Bagging and Weighing Scales—Automatic Bagging and Weighing Machine—Turpentine Still—Flour Packer.

SECTION XIX.

TEXTILE AND MANUFACTURING DEVICES, ETC.

Pattern Burring Machine—Cotton-Seed Hulling Machine—Cotton Bat Compressor and Condenser—Cocoanut-Paring Machine—Flock Grinding Machine—Flax-Scutching Machine—Multiple-Strand Cordage Machine—Paper Enameling Machine—Cordage-Making Machine—Three-strand Cordage Machine—Thirty-two-Strand Cordage Machine—Flocking Machine—Electric Cloth Cutter—Quarter Sawing of Lumber—Evolution of the Lag Screw—Porcelain Molding Machine—Diamond Cutting—Diamond Crusher and Mortar—Diamond Hand Tools and Drills—Combination Press—Artificial Flower-Branching Machine.

SECTION XX.

ENGINEERING AND CONSTRUCTION, ETC.

Four-Spool Hoisting Engine—Disintegrator—Foundry Construction—Excavator and Rotary Screen—Universal Pocket Level—Adjustable Beam Clamp—Gravity Elevator—Portable Concrete Mixer—Concrete Mixer—Trench Brace— Types of Machine-Shop Construction—Wood Preservation Apparatus—Wire-Guy Gripper—Timer Creosoting Apparatus—Electrically Driven Hammer— Duplex Rolling Lift Bridge—Balanced Swing Bridge—Fall Rope Cable Carrier —Crib Dam—Counterbalanced Drawbridge—Earth Embankment—High Structures—Transfer Bridge—Gigantic Wheel—Moving Platform—Traveling Stairway or Ramp.

SECTION XXI.

MISCELLANEOUS DEVICES.

Portable Saw—Stump-Pulling Machine—Motor Roller-Disk Plow—Automobile Plow—Reversible Plow—Tethering Hook—Fountain Wash Boiler—Potato-Washing Machine—Potato-Rasping Machine—Paris Green Duster—Automobile Mowing Machine—Modern Two-Horse Mower—Cream Separator—Refrigeration—Model· Cold-Storage House—Modern Grain Harvester—Compound Thresher—Refuse Crematory—Conical Charcoal Kiln—Coking Oven —Destructor Furnace—Life-Saving Net—Remington Typewriter—United States Army and Navy Guns—United States Magazine Rifle—Breech-Block Mechanism—Magazine Pistol—Artificial Ankle—Artificial Leg.

SECTION XXII.

DRAUGHTING DEVICES.

Geometrical Pen—Ellipsograph—The Campylograph.

SECTION XXIII.

PERPETUAL MOTION.

Perpetual Motion—The Inventor's Paradox—The Prevailing Type—Marquis of Worcester—Folding-Arm Type—Chain Wheel—Magnetism and Gravity— The Pick-up-Ball Type—The Ball-Carrying Belt—Ferguson's Type to prove its impossibility—Revolving Tubes and Balls—Geared Motive Power—The Differential Hydrostatic Wheel—The Lever Type—The Rocking Beam—Tilting Tray and Ball—The Rolling Ring which did not Roll—Differential Water Wheel—The Gear Problem—Mercurial Wheel—The Air-bag Problem—Air Transfer in Submerged Wheel—Extending weights and water transfer—The

SECTION I.

MECHANICAL POWER, LEVER.

MECHANICAL APPLIANCES

AND

NOVELTIES OF CONSTRUCTION.

Section I.

MECHANICAL POWER, LEVER.

---

1. LEVER IN A DRAUGHT EQUALIZER for four horses. This equalizer consists of a doubletree having singletrees, a bar pivoted

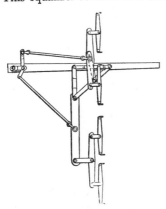

at one end to a lateral frame on the pole and connected at its outer end to the doubletree, a crossbar pivoted to the rear end of the pole being connected at one end by a rod connected at its other end to the bar pivoted to the lateral frame on the pole. The singletrees on the opposite side of the pole are pivoted to the end of a bar extending across the pole and pivoted to the crossbar. By this construction the draught of the horses secured to all the singletrees will be equalized, the doubletree on the pole being permitted to have a movement backward and forward on the end of a bar which is free to swing beneath the raised portion of a strap secured to the pole.

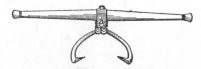

2. TIMBER OR LOG GRAPPLE. A handy device for carrying heavy timbers, joists, railroad ties, telegraph poles, etc.

3. LEVER EQUALIZER FOR SULKY PLOWS. Two jaws, forming a double clevis, attached to the front end of the beam

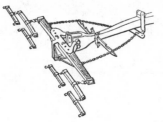

of the plow, and so arranged that by means of a series of holes in the jaws the plow may be regulated to run at a greater or less depth, and also to cut a furrow of any desired width.

Two levers of different lengths, to which the draught eveners of the team are secured, are pivoted one on either side of the jaws, and are connected by a chain that passes around a sheave secured on the under side of the drawbar. By this means the draught is equalized between the two beams. Swinging arms, pivoted to the sides of the beam, sustain the chains and hold them so as to draw straight from the equalizing levers.

4. LEVER EQUALIZER FOR THREE HORSES on single pole. The arms, A C, are fastened to opposite sides of the tongue,

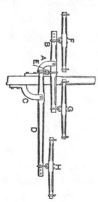

and the pivots in their ends are at equal distances from the tongue. To the end of the arm, A, is pivoted a doubletree, B, to one end of which a singletree, G, is held, and to the opposite end a singletree, F, is held adjustably by a pin passed through one of a series of holes in the end of the doubletree. The doubletree is pivoted about two-fifths of its length from the outer end. To the end of the arm, C, is pivoted a doubletree, D, on the outer end of which a singletree, H, is held by a pin through one of a row of holes. The inner end of this doubletree is connected by loops, E, with the middle of the doubletree, B. The doubletree, D, is pivoted about one-third of its length from its inner end. By means of the holes in the ends of the two doubletrees the leverage can be varied to suit conditions. The direct draught of the tongue is in the center of the two draught points.

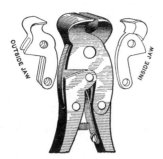

5. LEVER NIPPERS. A labor-saving device in the hands of the wire worker. Its lever advantage is readily seen by inspection of the detailed parts as a compound lever, which doubles the cutting power of the nippers.

5A. THREE-HORSE EVENER. This three-horse evener differs from the ordinary kind in that the extension parts are used to hitch the third horse in front of the tongue. It is made by using the doubletree for a two-horse team and, in place of the single-trees, use the pieces shown. Each of these pieces is a little longer than one-half the length of the doubletree. Each piece is fastened to the doubletree in place of a singletree at a point one-third the way from one end. The singletrees are fastened to the short end of the pieces, and the longer ends

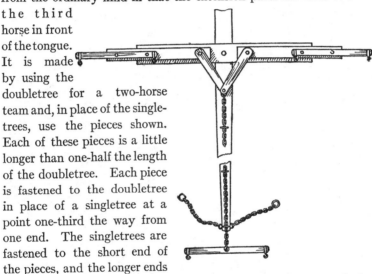

of the two are joined with two bars and a clevis which is attached to a chain running on the under side of the tongue. The chain is kept from sagging by several hooks or rings fastened in the wood. This evener prevents the lead horse from pulling the greater part of the load. The engraving shows the under side of the tongue.

SECTION II.

TRANSMISSION OF POWER.

Section II.

TRANSMISSION OF POWER.

6. UNIVERSAL SCREW DRIVER. The handle has a ratchet socket in which the three-point blade may be inserted for greater power or to accommodate special conditions.

7. A section showing the ratchet and pawl socket for holding the square shank of the blade for corner work.

8. QUICK COUPLING for sewer rods. Makes a smooth continuous rod that can not uncouple while in use.

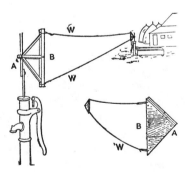

9. TRANSMISSION of power by wire rope and anchored levers. The braced tee pieces A B, with their arms connected with a distant rocker by the wire W W, pivoted to the windmill frame and to the crank rod at A, make a very effective method of operating a pump at a distance. A strong fence wire is sufficient for a house pump and may be supported on rollers for long distances.

23

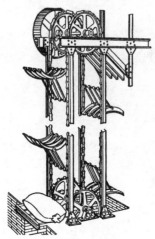

10. BAG ELEVATOR. The bags are delivered from a car door on to a grating through which the forked hands of the elevator picks them up and discharges upon an inclined chute grating to slide to a horizontal carrier from which they are deposited at any desired place. The forked hands are braced loosely to the sprocket chains, which are guided in grooved posts, so that there is no sag to the forks when the load is on.

11. HORIZONTAL CONVEYOR. Receives the bags from the elevator (Fig. 10) and deposits them along a warehouse floor

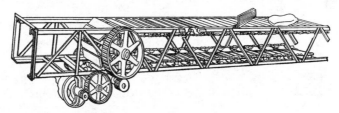

by dumping them off the side at places where the inclined guide board is set.

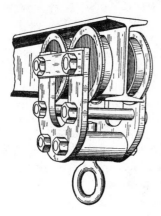

12. I BEAM TROLLEY. A simple and effective apparatus with a chain tackle for setting heavy work in lathes and moving light articles in shops.

The I beam makes a most convenient outrigger from the front of a warehouse or factory for the transfer of goods to and from trucks.

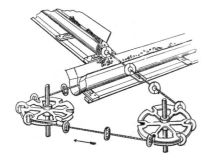

13. TWO WAY CON-VEYOR. Method by which a rope and disk conveyor can be made to change its direction. For grain, gravel, sand, clay, and other loose material.

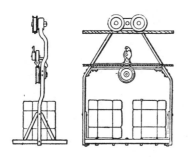

14. ROPE TRAMWAY CAR-RIAGE. Bleichert System. The upper rope is the bearing cable and trolley, the lower one is the hauling rope with the grip device attached to the car frame. The pull of a lanyard starts or stops the car.

15. Shows a side view of the open car frame and grip cam.

16. FRICTION PULLEY. The central hub A, which is keyed to the shaft, is turned up to form a bearing for the pulley and for the

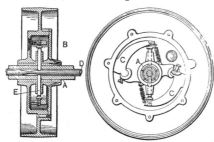

cover B fastened over the circular chamber in the pulley. The gripping dogs or levers C—hung at the ends of the arms on hub A—are finished at one end to fit the friction surface in the pulley chamber. The countershaft is drilled out to receive the hardened rod D, which is connected to the shipper. As the rod is moved in the shaft by the shipper, a double wedge—formed on the rod—forces out the two pins E, and these pins press the gripping levers tight against the friction surface. When the rod is moved in the opposite direction the springs force the pins toward the center and re-lease the levers. A screw plug at the back of the chamber can be re-moved and the pins E adjusted to give the gripping levers the desired pressure.

17. A section showing the pins and wedge rod.

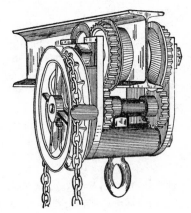

18. GEARED I BEAM TROLLEY. Designed for moving heavy articles on overhead I beam railways in factories. The trolley wheels are geared to a driving shaft with sprocket wheel and chain, the lift being an ordinary tackle, not shown.

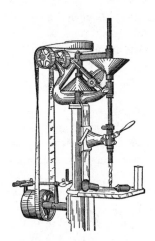

19. VARIABLE POWER AND SPEED with friction cone pulleys and traversing pulley as applied to a drilling machine. The transmitting roller is pivoted in a frame that slides on a side bar and is clamped by a screw at the position required for the desired speed. See No. 106, 1st vol., for a frictionless form of transmitting roller.

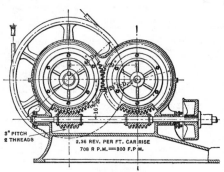

20. WORM GEAR ELEVATOR. Sprague type. The double worm and gear serves the purpose of balancing the thrust of the driving shaft and is also a means of safety from breakage of teeth. The wear on the worms and gear is also much lessened by their duplication.

21. CASH CARRIER. To the upper surface of the car are secured uprights, in which are journaled the axles of grooved wheels

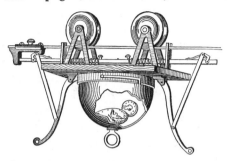

running upon the overhead wire or track. In other uprights is held a rod on which are placed two coiled springs, so arranged that the rod acts as a double buffer to the carrier, each of its ends being adapted to strike a stop block, two of which are attached to the wire, one at each end. Near each end of the bar is a pawl, acted upon by a spring which lifts its free end so it will automatically engage with a lip formed on the stop block for holding the car stationary when it reaches either end of its trip. The pawls are disconnected and the car started by means of levers pivoted to the frame and connected with the pawls.

22. VARIABLE SPEED DEVICE. The wheels A and A' are each made up of two disks mounted on a shaft and carrying between

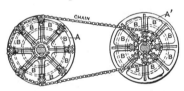

them small toothed pinions, B B B, and B' B' B', which are mounted on roller clutches. The bearings for the pinions are arranged to move radially in the slots shown in the plates, so that the diameter around which the chain must wrap, may be lengthened or shortened at will. If the pinions of one wheel or drum are moved radially outward, those of the mating drum must be moved inward, and *vice versa.*

The pinions are moved radially by means of two scroll plates for each sprocket, the spiral slots of which engage the bearings of the pinions and move them in the same manner as the jaws of a scroll chuck are operated. The manner in which the scroll plates are turned, to effect changes in diameter of driving and driven gears, is accomplished by the simultaneous moving in or out of two flat racks lying in slots cut in the sprocket shafts.

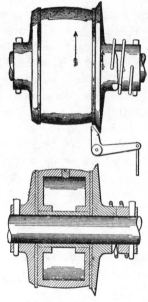

23. FRICTION PULLEY. The flange at the left is fast on the shaft, while the flange at the right is loose. On the end of the hub of the latter flange teeth are cut, the surfaces between the teeth being helical, as shown. The fixed collar at the right is milled to correspond. The spring secured to the loose flange and the collar is always under tension and tends to rotate the flange in the direction in which the belt travels. As the flange turns on the shaft it is forced against the running pulley and is then turned by the friction until the pulley is clamped fast between the two friction surfaces, when the pulley, flanges, and shaft all rotate together. To release the pulley, the brake —shown just below the right-hand friction disk—is brought against the angular face of the flange. This holds the flange back, but the collar still turns ahead with the shaft, thus removing the end pressure on the friction and releasing the pulley.

24. A section showing the details of construction.

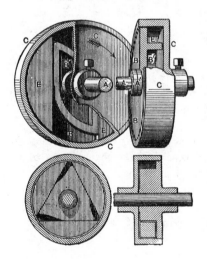

25. PANEL CLUTCHES. Simpson type. A silent clutch that prevents back movement and takes up a forward motion without the jerk of a ratchet and pawl. Useful on agricultural implements, sewing-machines, etc.

26. The under figure represents a triangular quick-action panel applied on the same principle as the other against three friction segments. Plan and section.

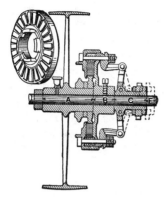

27. FRICTION PULLEY. Upon the hub of the pulley is keyed a collar with lugs on one side which engage with lugs on the friction disk ; this causes the disk that fits loosely on the hub of the pulley to revolve with the pulley, and at the same time leaves the disk free to vibrate sideways if necessary. The advantage of the disk being loosely connected with the pulley in this manner, will be appreciated in case the pulley should become worn loose on the shaft.

28. VISCOSIMETER. An instrument for measuring the viscosity of liquids, or the resistance which a liquid offers to flowing or a quick change of state. The liquid to be tested is placed in the reservoir in which is a paddle-shaped agitator or wheel. The shaft of this wheel is run by a train of gears actuated by a drum, which is caused to revolve by means of a weight and cord as shown. At the upper end of the shaft is a worm and worm wheel and on the shaft of the worm wheel is a pointer, which passes over the face of a dial, by which the speed of the paddle or agitator can be ascertained. The

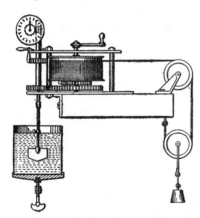

weight is first drawn up by means of the crank on top of the drum. The liquid is poured into the reservoir and the latter raised to the proper point to give the paddle wheel the proper submersion. The trip on the paddle shaft just below the dial is then thrown out, when the drum and weight start the paddle wheel revolving. The viscosity of the liquid, with reference to some other liquid taken as a standard, is then determined by noting the indications of the pointer on the dial. The number of revolutions of the pointer or the number of divisions passed over in a given time, compared to the reading when testing another liquid, indicates the relative viscosity.

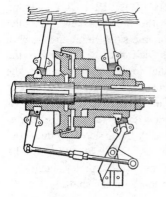

29. POSITIVE COMBINATION CLUTCH. The first motion of the clutch handle brings the friction cones into contact; a further push of the handle moves the teeth of the clutch into contact and prevents slipping of the friction cones. The bell-crank arm on the handle holds the clutch fast in its locked position.

30. PNEUMATIC BELT SHIPPER. The device consists simply of a small air cylinder with a piston travel, such as will give the

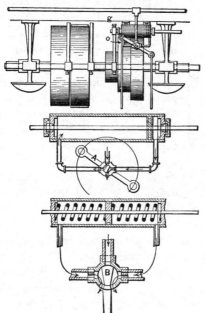

belt the proper throw; the cylinder is piped from each end to a two-way cock, the plug of which has a bar with a looped cord within reach of the operator. Attached to the piston is an arm o, which extends down to the bar carrying the shifter forks—air does the rest. Section 31 explains the whole thing for a belt requiring but one movement.

Machines having a backing belt are provided for by the arrangement shown in section 32, in which the piston and belt shifter are held in a central position by two coiled springs when the belts are on the loose pulleys. The springs are compressed, and their resistance is easily overcome, when air is admitted to the opposite ends of the cylinder, which action will put the belt on either the forward or backing pulleys.

33. ACOUSTIC TELEPHONE. The mouthpiece, *a*, has a central aperture for the passage of sound waves to the diaphragm, *c*,

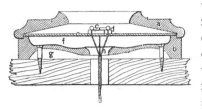

whose edges are secured within a rabbet of the mouthpiece. The diaphragm is about 7 inches in diameter and is made of spruce wood, which possesses great sonorousness combined with strength sufficient to sustain the tension of the line wire. The mouthpiece and diaphragm are held to the wall on a bed piece, *b*, by the tension of the line wire. The bed piece is recessed at both sides, *fg*, and centrally apertured for the passage of threads connecting the line wire to the diaphragm. The front recess, *f*, affords a space between the diaphragm and the center of the bed piece for free action of the diaphragm, promoting clearness of enunciation when the instrument is used as a receiver, and the rear recess, *g*, secures a small marginal support for the transmitter, thereby avoiding a large contact with the wall and preventing excessive vibration.

To avoid indistinct articulation and the ringing sounds common to acoustic telephones, the line wire is connected to the diaphragm by silk cords, which are twisted about the end of the wire to obtain a firm connection therewith, and which diverge into three or more strands that are secured to a metal ring, *c*, between which and the diaphragm a rubber or leather ring, *d*, is interposed. The line wire is made of strands twisted together and coated with varnish to bind them and prevent them rubbing upon one another.

34. ACOUSTIC TELEPHONE. Wire suspension to facilitate

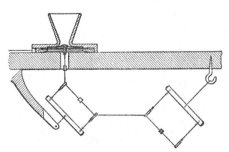

transmission of sound by making the angular turns at about 45°, with rubber straps wired to a yoke of wood as shown. Small dampers of rubber or leather are wired or tied on to the main wire between nodal points to prevent undue vibration and change of tone by wind or rain. The mouthpiece, if of metal, should be grounded, to prevent electric sparks.

35. ACOUSTIC TELEPHONE.   The front board, A, of the box is provided with a central aperture.   The diaphragm, *b*, is stretched over

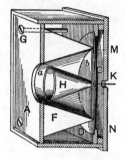

the central opening of a board, D, which has strengthening ribs on its under side and along the edges.   An annular block of wood, F, whose thickness decreases from the top toward the bottom on the inner as well as the outer side, is placed between the front board and the diaphragm.   The upper opening of the block coincides with the central opening of the front board, and the bottom opening is smaller than the opening in the board, D.   The bottom edge of the block is pressed upon the diaphragm by bolts, G.   In the central opening of the block is a funnel-shaped vessel, H, held in place by wires, *a b*, at the top and bottom, which hold the lower end of the funnel a short distance from the diaphragm.   A button is fastened to the middle of the diaphragm, to which is fastened the wire, K.   The funnel concentrates the sound waves and guides them to the diaphragm, thus causing strong and distinct vibrations that reproduce the words very plainly.   The diaphragm is formed of alternate layers of skin and a textile fabric, or of hard rubber about $1/_{16}$ inch thick.

36. ACOUSTIC TELEPHONE.   The characteristics of this tele-

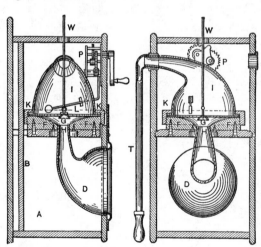

phone are a curved mouthpiece, D, a vibrating chamber, I, with an ear tube, T, for returning the vibrations to the ear without moving the head in conversation. It also has attached a clock-gear vibrator, which makes a loud call by the hammer strokes on the diaphragm. The mouthpiece, D, and the resonator, I, may be made of metal, hard rubber, wood, or *papier-maché* as convenient, from 3 to 5 inches in diameter.   **37.** Cross section.

SECTION III.

MEASUREMENT OF POWER, SPRINGS.

Section III.

MEASUREMENT OF POWER, SPRINGS.

---

38. REGISTERING WIND VANE. The wind vane, *a a*, 15 feet above the roof of the building, the vertical shaft, *b b*, which is

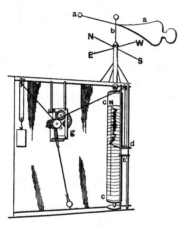

fastened to and turns with the vane, and gives corresponding movements to the cylinder, *c c*, round which the register paper is fastened. The paper is divided vertically into 24 parts for the hours of the day and its circumference on the drum into 4 divisions, N, E, S, W, for the quarters of the wind. At *d*, is a pencil tube attached to the top of the weight, *e*, and the pencil bears lightly against the paper by an India rubber elastic spring. A clock, *g*, permits this weight to descend to the bottom of the cylinder in twenty-four hours, when the cylinder is taken out, and the register slipped off. Another one is put on by pasting one of the edges of the paper and letting it overlap the other edge round the cylinder.

39. ANEMOMETER. Robinson type. For measuring the velocity of the wind. The worm screw runs in one of a train of geared index wheels carrying pointers. The dials are marked for direct reading of from 0.1 mile, 1 mile, 10 miles, 100 miles, 1,000 miles ; so that the difference of two readings and the time gives a ready solution of the wind velocity.

40. ELECTRIC SIGNAL ANEMOMETER.   Consists of four hollow hemispheres or cups fixed to the ends of two horizontal rods A crossing each other at right angles and supported on a vertical shaft D

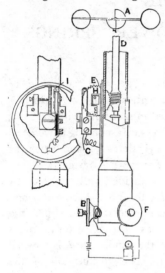

whose worm thread is geared into the wheel H.   On the face of wheel H are beveled projecting pins E which engage with a beveled projection on the spring I. The pins are so arranged on wheel H as to momentarily close circuit through a single stroke bell for every twenty-five revolutions of the cup arms A.   The electric circuit starts at the insulated binding-post B, goes through wire to the insulated spring C, where it is completed to spring I to frame of instrument and the uninsulated binding-post F. From binding-posts B and F, where start the line wires, the circuit is completed through bell and battery.

41. METALLIC THERMOMETER.   The instrument is provided with two series of hard rubber pulleys mounted on studs projecting from a board.   A fine brass wire

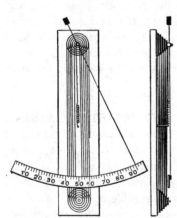

(No. 32) attached to the board at one end passes around the successive pulleys of the upper and lower series in alternation, the last end being connected with one end of a spiral spring, which is strong enough to keep the wire taut without stretching it.   The other end of the spring is attached to a stud projecting from the board.   The pulleys are of different diameters, so that each series forms a cone.   By this construction the wire of one convolution is prevented from covering the wire of the next.

· The last pulley of the upper series is provided with a boss, to which is attached a counterbalanced index.   A curved scale is supported behind the index by posts projecting from the board.

42. WIND FORCE REGISTER. The metal drum, *a a*, is made of tin plate ; it is two feet in height and one foot in diameter, suspended

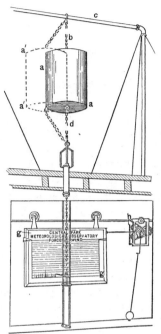

by a chain, *b*, to a strong support, *c*, on the roof of the Observatory ; its lower end is connected by a chain and guide rod, *d*, with a spiral spring at the bottom of the case. On the top of this spring is a pencil; it bears against the sheet of paper, *g g*, fastened to a board drawn aside by a clock. When the wind blows, the tin cylinder is forced into some other position, as is shown in the dotted figure, *a' a'*, and the pencil is drawn upward. The more violent the wind the further will the tin cylinder be pushed aside, and the higher the pencil drawn. The force necessary to draw the pencil upward to a given point is determined by direct experiment, and expressed in pounds weight upon a square foot. The direction from which the wind blows makes no difference, as it always has the same surface to press upon.

43. RECORDING WIND METER. On the ends of a cross supported by a vertical shaft several feet above the roof of the building,

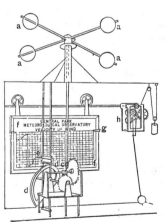

are four hemispherical copper cups. These, whatever may be the direction of the wind, are caused to turn round with a speed, as has been determined by experiment, of about one-third the velocity of the wind.

To the lower end of the shaft thus made to revolve by the cups is attached an endless screw connecting with a train of wheels, which move a cam. The wheels are so arranged that one turn of the cam answers to 15 miles in the movement of the wind. A pencil which rests on the edge of the cam, and bears lightly against a surface, is carried from the bottom to the top of the paper by each revolution of the cam. It should be understood that the paper is attached to a board drawn aside by a clock at the rate of half an inch an hour. The number of times that the pencil moves from the bottom to the top of the paper, multiplied by 15, gives the number of miles that the wind has moved in an hour or day.

44. RECORDING BAROMETER. The tube marked A B is of glass ; the upper part is of a larger diameter than the stem, A being

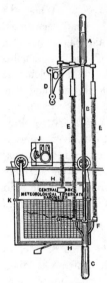

$^3/_4$ of an inch internal diameter and 10 inches long, while the stem, B, is $^1/_8$ of an inch bore and 26 inches long. The total length of the tube is therefore 36 inches. The reservoir, C, is suspended from a brass frame, D, fastened to the back of the case. This frame also holds the upper ends of the steel springs, E, E. The glass reservoir, C, is of the same diameter and length as the upper part of the tube, A ; on its open end is turned a flange to hold it in a brass frame, F, to which are fastened the lower ends of the steel springs, E, E ; it also carries an ink pencil, G, that touches the ruled paper on the board, H, H, which is drawn aside by the clock, J.

The springs for weighing the reservoir are made of steel wire, No. 22 English wire gauge, closely wound round a mandrel, $^1/_2$ inch in diameter and 10 inches long, on which they are tempered hard and afterward lowered to a suitable temper by being dipped in oil and ignited two or three times, the burned oil forming a japan that prevents them from rusting in damp weather.

Ink pencils of the barometer and other instruments are made by drawing narrow glass tubing to a fine point, which lightly touches the paper register, leaving a mark of red ink that has been diluted with about one-quarter of its volume of glycerin. The glycerin prevents the ink from drying too rapidly.

To receive the atmospheric fluctuations a suitable ruled paper is fastened by means of small brass clamps, K K, to the board, H H, which is hung by rollers to the thick steel rod fastened to the sides of the case, on which the paper is carried from right to left by the clock, J, at the rate of $^1/_2$ an inch per hour, by means of the pulley on the hour arbor of the clock.

45. REGISTERING AIR THERMOMETER. A glass tube bent into a circle with an air bulb at the top nicely balanced by coun-

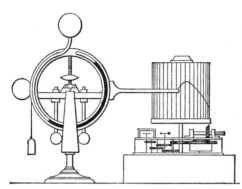

terweights on a knife edge. Part of the tube, as shown by the heavy black line, is filled with mercury. The expansion and contraction of the air in the bulb end moves the mercury and carries the pointer up or down by change of gravity. A cylinder moved by clockwork receives the record.

46. METALLIC THERMOMETER. A series of rods or strips of zinc and iron alternating, with their alternate ends fastened together,

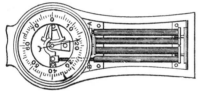

so that the first rod shall be zinc and the last iron, which should extend to a curved lever that operates a sector and small pinion to which the index hand is attached. The dial graduation to be made by comparison with a standard thermometer.

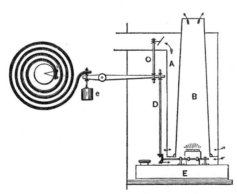

47. THERMOSTAT. For an incubator. A coil of zinc and steel ribbon fastened together by rivets or solder is fixed at its center end to a block in the incubator chamber and attached by a lever to a damper and to a lever on the wick gear, and controls both air inlet and flame. The zinc should be thicker than the steel and on the outside of the coil. *e*, counterweight for balancing lever and connections ; A, air inlet cylinder ; B, lamp chimney, all of sheet iron ; O, damper rod ; D, wick rod.

48. METALLIC THERMOMETER. The instrument depends for its operation on the difference between the expansion of brass and

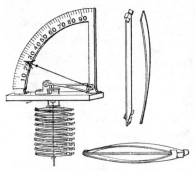

steel. The linear expansion of brass is greater than that of steel, so that when a curved bar of brass is confined at the ends by a straight bar of steel, the brass bar will elongate more than the steel bar when both are heated, and will in consequence become more convex.

At the right are shown two bars, the straight one being of steel, the curved one of brass. The steel bar is slit for a short distance in two places at each end, and the ears thus formed are bent in opposite directions to form abutments for the ends of the curved brass bars, two brass bars being held by a single steel bar, thus forming a compound bar, as shown below. Each compound bar is drilled through at the center. Ten or more such compound bars are strung together loosely upon a rod, which is secured to a fixed support. A stirrup formed of two rods and two crosspieces rests upon the upper compound bar and passes upward through the support. Above the support it is connected by a link with a sector lever which engages a pinion on the pivot of the index.

49. Straight steel and curved brass bar.

50. The compound bar.

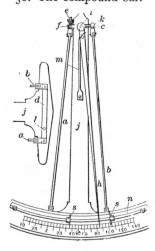

51. MAXIMUM AND MINIMUM RECORDING THERMOMETER. *a, b,* two strips or rods of hard rubber attached to the index arm *k,* and the adjusting post *f.* The lower ends pivoted to the lever *d, l,* shown on the side figure. *m,* a spring to take up looseness. *j,* a wooden frame. *s, s,* two light flanged sleeves sliding freely on the bent rod *n,* which is fixed above the index scale. The scale to be made by comparison with a mercurial thermometer. *e,* a nut for adjusting the position of the index.

52. SUNSHINE RECORDING THERMOMETER. Extend-
ing from the instrument in the room through the roof is an iron pipe.

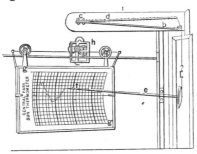

On its upper end it carries an
iron bar, *b*, to which is fastened
firmly, at *c*, the metallic thermometer bar, *d;* from the loose
end of this bar a fine wire
descends over a guide pulley
down the inside of the pipe to
the lever, *e*, in the case in the
room below.   On the long end
of the lever is an ink pencil, *f*,
that records the motions of the thermometer bar on the register sheet,
g g, which is fastened by means of two small clamps to the board that
is carried sideways by the clock, *h*.   Over the metallic thermometer
bar above the roof is a glass shade, *i*, protecting it from the weather,
or covered with a louver box when used for recording temperature only.

On the register paper, *g g*, are shown the fluctuations for the day.
In this instance many clouds have passed between the sun and instrument.   If the curve had been without oscillations it would show that
the day had been clear.

53. CENTRIFUGAL SPEED INDICATOR. Gravity of a
colored fluid in the central and outer tubes is varied by the centrif-

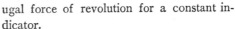

ugal force of revolution for a constant indicator.

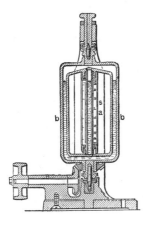

The machine consists of three tubes, *a, b,*
b¹, which connect freely with one another,
and are mounted vertically between conical
centers.   The tubes are sealed air tight so
that no liquid can escape or be added.   A
scale is placed on a standard opposite the
central glass tube, and is graduated to correspond to various speeds.   When the apparatus is set in motion, the level of the colored liquids falls in the central tube *a*, and
rises in *b*, and *b¹*; by comparing the level of
the liquid in *a* with the scale, the speed can
be read off.

54. HYGROSCOPE. In the instrument a strand, H, of hair, deprived of all fat, is secured at its upper end at f, and at its lower end to a crank, k, carried by the shorter and heavier arm,

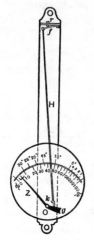

g, of an angle-lever pivoted at O. The longer and lighter arm of the lever serves as a pointer and terminates in a trident, Z. Hair has the property of expanding or lengthening with an increase of relative moisture and of contracting with a decrease in relative moisture. Since the strand of hair is constantly under the tension imposed by the weight of the arm, g, an increase or decrease of relative moisture and a corresponding expansion or contraction of the hair will be accompanied by a movement of the pointer, Z, which plays over a double scale. The central point of the arm, Z, indicates on the lower scale (graduated from 0 to 100) the relative moisture.

Hygrometers which are employed at no great elevations are influenced by the moisture of the soil after a heavy fall of dew or rain. For this reason five, eight, ten, or fifteen per cent. must be deducted from the percentage indicated by the pointer, for light rain (snow, fog), moderate rains, and heavy, continuous rains.

55. PRONY BRAKE. The lever arm is pivoted at A. The band carrying the brake blocks is connected to the lever at D and B. The

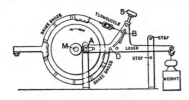

brake blocks are hollow and provided with internal water circulation for cooling. The faces of the brake shoes are smeared with tallow, and no water is allowed on the friction surfaces. The block B, to which the band is attached, moves in a curved slot, being controlled by the screw and handwheel S. A turnbuckle is provided in the band for tightening the grip of the blocks. A very close regulation may be obtained by means of the various adjustments, since the coefficient of friction fluctuates very slightly owing to excellent lubrication and absence of water from the friction surfaces. It is necessary that the center M of the shaft, the pivot at A and the point of attachment of the weight to the lever all be in the same straight line parallel to the ground line.

56. TRANSMISSION DYNAMOMETER. The motor acts directly upon the axle of the wheel, A, in the direction shown by the

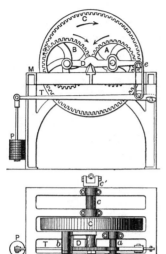

arrow, and this wheel carries along the intermediate one, B, which transmits motion to the inner-toothed wheel, C. The latter is connected with the machine to be experimented upon by the axle, c, and the Cardan joint, c'.

The axles, a and c, revolve in bearings fixed to the frame, M, but the axle of the wheel, B, revolves in a bush which is carried by a beam whose fixed axis passes exactly through the contact of the circumferences of the wheels, A and B. The result of this is, that the momentum of the force exerted by the wheel, A, upon B, is null with respect to the edge of the knife-blade upon which the beam oscillates, and that, consequently, such force has no tendency to move the beam in one direction more than in another. The beam, then, is only influenced by the resistance that the wheel, C, offers to the motion of the wheel, B ; and it is such resistance that, by a system of levers in a ratio of 1 to 10, is measured by means of the weight, P.

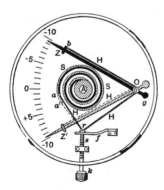

57. THERMOHYDROSCOPE. The instrument comprises essentially a double spiral, S, of zinc and iron and a prepared strand, H, of hair, extending from the end, a, of the spiral, through an eye, r, over the roller, O, to the end of the index, Z. The eye, r, is carried by the spring, f, and is raised or lowered by means of the set-screw, s. If the eye be lowered, the strand of hair is subjected to tension, and the index, Z, thereby raised. If the eye be raised, the strand of hair is slackened, and the index falls by its own weight. In this manner the index of the instrument is adjusted.

The spiral, S, operates in the same manner as the spiral of a thermometer. When the temperature rises the spiral curves inwardly, so that its free end, *a*, moves downwardly to *a'*. The tension of the strand, H, is thereby diminished, and the index falls. Since the relative moisture has remained the same, the length of the strand is not changed. While the polymeter points constantly to fifty per cent., the thermohygroscope, through the falling of its index, shows that the temperature has raised, and with a uniform temperature, a change of the index indicates change of the relative moisture.

POWER OF SPRINGS.

Reference to letters of formula :

P. Maximum in pounds.

B. Breadth of spring in inches.

H. Thickness of spring in inches.

L. Length of spring in inches.

F. Deflection of spring in inches.

R. Radius of helical springs or points at which load is applied.

S = max. stress.

= 100,000 lbs. per sq. in. for elliptical springs, ⎱ Steel.
= 80,000 " " helical " ⎰
= 14,500 " " " " Brass.

E = modulus of elasticity.

= 31,500,000 lbs. per sq. in. (steel).
= 15,000,000 " " (brass).

G = modulus of elasticity for torsion.

G = $\frac{2}{5}$ E = 12,600,000 lbs. per sq. in. (steel).
= 6,000,000 " " (brass).

Best work of spring is at one-half its maximum load as per formula·

58. RECTANGULAR SPRING. Fast at one end load at the other end.

Max. load.	Deflection.	Flexibility.
$P = \dfrac{S\,B\,H^2}{6\,L}$	$F = \dfrac{6\,P\,L^3}{E\,B\,H^3}$	$\dfrac{F}{L} = \dfrac{S\,L}{E\,H}$

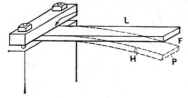

If spring is triangular in breadth and of equal thickness, use above formula, in which B = breadth at base or widest part.

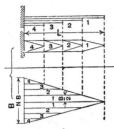

59. COMPOUND TRIANGULAR SPRING, or more than one leaf of either form in the cuts.

Max. load.	Deflection.	Flexibility.
For a single and double elliptical spring the max. load = 2 P. $$P = \dfrac{S N B H}{6 L} = \dfrac{S B^1 H^2}{6 L}$$ N = number of leaves.	Deflection of a double elliptical spring = 2 F. $$F = \dfrac{6 P L^3}{E N B H^3}$$	$$\dfrac{F}{L} = \dfrac{S L}{E H}$$

Ends of leaves tapered to $\dfrac{H}{2}$

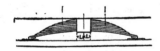
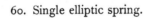

60. Single elliptic spring.

61. Double elliptic spring.

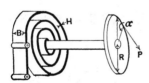

62. VOLUTE OR SPIRAL SPRING, flat. P = power applied at end of arm, R. Distance a = flexure of arm.

Max. load.	Deflection.	Flexibility.
$$P = \dfrac{S B H}{6 R}$$ R = radius of P.	$$F = R\, a = \dfrac{12\, P L R^2}{E B H^3}$$	$$\dfrac{F}{R} = \dfrac{2 S L}{E H}$$

L, total length of spring in action.

63. HELICAL SPRING, flat. $P =$ power applied at end of arm, R. Distance $a =$ flexure of arm.

Max. load.	Deflection.	Flexibility.
For square $B = H.$ $P = \dfrac{S\,B\,H^2}{6\,R}$ $R =$ radius.	$F = R\,a = \dfrac{12\,P\,L\,R^2}{E\,B\,H^3}$	$\dfrac{F}{R} = \dfrac{2\,S\,L}{E\,H}$

$L =$ total length of spring in action.

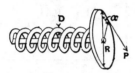

64. HELICAL SPRING, round. $P =$ power applied at end of arm, R. Distance $a =$ flexure of arm.

Max. load.	Deflection.	Flexibility.
$P = \dfrac{S\,\pi\,D^3}{32\,R}$	$F = R\,a = \dfrac{64\,P\,L\,R^2}{\pi\,E\,D^4}$	$\dfrac{F}{R} = \dfrac{2\,S\,L}{E\,D}$

$L =$ total length of spring in action.
$\pi = 3.1416.$ $D =$ diameter of steel.

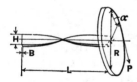

65. STRAIGHT TORSION SPRING, flat.

Max. load.	Deflection.	Flexibility.
$P = \dfrac{S\,B^2\,H^2}{3\,R\sqrt{B^2+H^2}}$ $H > B,$ nearly. $P = \dfrac{S\,B^2\,H^2}{3\,R\,[0.4\,B+0.96\,H]}$	$F = R\,a$ $= \dfrac{3\,P\,R^2\,L\,[B^2+H^2]}{G\,B^3\,H^3}$	$\dfrac{F}{R} = \dfrac{S\,L\sqrt{B^2+H^2}}{G\,B\,H}$

H, breadth of spring.   B, thickness.   G, modulus $= \frac{2}{5}\,E.$

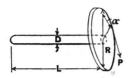

66. STRAIGHT TORSION SPRING,

round. P = power applied at end of arm, R. Distance a = flexure of arm.

Max. load.	Deflection.	Flexibility.
$P = \dfrac{S \pi D^3}{16 R}$ R = radius.	$F = R a = \dfrac{32 P R^2 L}{\pi G D^4}$	$\dfrac{F}{R} = \dfrac{2 S L}{G D}$

$\pi = 3.1416.$ D = diameter of steel.   G, modulus of elasticity for torsion.

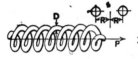

67. HELICAL TORSION SPRING,

round. To pull lengthwise of the helix.

Max. load.	Deflection.	Flexibility.
$P = \dfrac{S \pi D^3}{16 R}$	$F \quad \dfrac{2 R S L}{D G}$ $F = \dfrac{32 P R^2 L}{\pi G D^4}$	$\dfrac{F}{R} = \dfrac{2 S L}{G D}$

$\pi = 3.1416.$ D = diameter of steel.   G, modulus.

68. HELICAL TORSION SPRING,

flat. To pull lengthwise of the helix.

Max. load.	Deflection.	Flexibility.
$P = \dfrac{S B^2 H^2}{3 R \sqrt{B^2 + H^2}}$ $H > B$, nearly. $P = \dfrac{S B^2 H^2}{3 R [0.4 B + 0.96 H]}$	$F = \dfrac{3 P R^2 L [B^2 + H^2]}{G B^3 H^3}$	$\dfrac{F}{R} = \dfrac{S L \sqrt{B^2 + H^2}}{G B H}$

G, modulus.

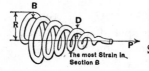

69. CONICAL SPIRAL TORSION SPRING. Round, to pull or push.

Max. load.	Deflection.	Flexibility.
$P = \dfrac{S \pi D^3}{16\,R}$	Nearly $F = \dfrac{16\,P\,R\,L}{\pi\,G\,D^4}$	$\dfrac{F}{R} = \dfrac{S\,L}{G\,D}$

$\pi = 3,1416.$ G, modulus.

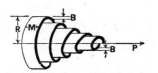

70. CONICAL SPIRAL TORSION SPRING, flat.
To pull or push.

Max. load.	Deflection.	Flexibility.
$P = \dfrac{S\,B^2\,H^2}{3\,R\,\sqrt{B^2 + H^2}}$ $H > B$, nearly. $P = \dfrac{S\,B^2\,H^2}{3\,R\,(0.4\,B + 0.96\,H)}$	Nearly $F = \dfrac{3\,P\,R^2\,L\,[B^2 + H^2]}{2\,G\,B^3\,H^3}$	$\dfrac{F}{R} = \dfrac{S\,L\,\sqrt{B^2 + H^2}}{2\,G\,B\,H}$

G, modulus.

71. BOLSTER SPRINGS, round.   For each spring, if double.

Max. load.	Deflection.	Flexibility.
$P = \dfrac{S \pi D^3}{16\,R}$	$F = \dfrac{2\,R\,S\,L}{D\,G}$ $F = \dfrac{32\,P\,R^2\,L}{\pi\,G\,D^4}$	$\dfrac{F}{R} = \dfrac{2\,S\,L}{G\,D}$

D = diameter of steel. $\pi = 3.1416.$ G, modulus.

72. COMPOUND BOLSTER SPRING. The value of each spring must be first obtained and all added for the compound spring.

	Max. load.	Deflection.	Flexibility.
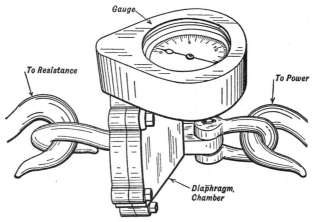	$P = \dfrac{S \pi D^3}{16 R}$	$F = \dfrac{2\,R\,S\,L}{D\,G}$ $F = \dfrac{32\,P\,R^2\,L}{\pi\,G\,D^4}$	$\dfrac{F}{R} = \dfrac{2\,S\,L}{G\,D}$

G, modulus.

72A. TRACTION DYNAMOMETER. The drawbar pull of any motor vehicle is determined by means of some form of traction dyna-

Gauge

To Resistance

To Power

Diaphragm. Chamber

mometer. This instrument is placed between the source of power or drawbar of the vehicle and some immovable body, such as a large tree. The vehicle is started and the maximum amount of pull is indicated in pounds upon the gage just before the vehicle driving-wheels slip or the engine stops. This dynamometer may also be used to indicate the tension in tow line or draft gear. As will be apparent, an instrument of this nature may be used very easily in making comparative tests between the tractive power or drawbar pull of various forms of engine and will also indicate the amount of draft needed to haul a wagon, pull a plough or do any other work.

SECTION IV.

GENERATION OF POWER, STEAM.

Section IV.

GENERATION OF POWER, STEAM.

73. INTERNALLY FIRED BOILER. Double corrugated tubular furnace with cylindrical shell and return tubes. A large volume of water and large water surface which insures steady steaming.

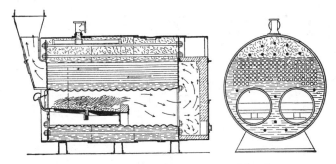

A liberal steam space and dry pipe prevents siphoning.
74. Cross section of boiler.
Continental type.

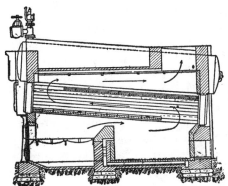

75. HEAT CIRCULATION in a Hein boiler. The two longitudinal firebrick partitions along the upper and lower tubes direct the heated gases in contact with the entire tube surface.

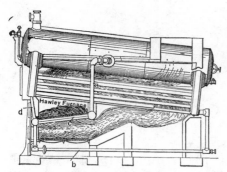

76. DOWN DRAUGHT BOILER FURNACE. Hawley type under a Hein boiler. *c*, tubular grate. *d*, tube connection between grate header and front drum. *b*, tube connection between grate header and rear drum with blow off.

a, uptake connection from each end of grate header to shell of boiler. This arrangement gives a rapid circulation in the grate-bars and prevents overheating.

77. TRIPLEX BOILER. Fanning type with down draught grate. Shells are filled with tubes and with no stays. Top shell used for

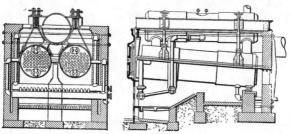

steam space. Gases of combustion pass under lower shells through the tubes and back between the three shells. Water line in upper shell. Large efficiency claimed.

78. Section through boiler.

79. WATER TUBE BOILER. Arranged for utilizing the heat

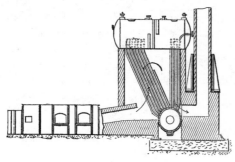

from a puddling furnace. A most efficient economizer of heat from any kind of furnace from which there is sufficient waste heat for generating steam. A diaphragm guides the heat to best advantage through the two sets of upright tubes.

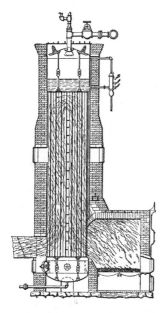

80. VERTICAL WATER TUBE BOILER. Wood type. A furnace chamber at one side; an upper and lower tube drum with the tubes banked in two sections, and a fire-tile partition extended nearly to the top of the tubes. The only provision for circulation is by the upward current in the fire side inducing a downward flow in the rear bank of tubes. Tubes are cleaned by steam jets, through doors in the walls of the setting.

81. FLASH COIL BOILER. Made with two open coils of iron pipe interlocked so that the central space is occupied by a useful steam-generating surface.

This form gives a large generating surface in a small space.

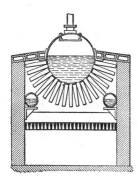

82. FINGER TUBE BOILER. The shell is made of thick tubing for small boilers or $\frac{1}{2}$ inch plate for larger size. The fingers are of short pieces of pipe welded at one end with a square head for a wrench and screwed into the shell, using ordinary pipe threads. The connecting tubes are not essential and may be omitted. An excellent boiler for amateur practice.

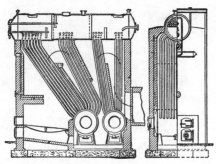

83. DUPLEX WATER TUBE BOILER. One of the many forms of water tube boilers now coming into general use in which great economy in evaporative power and space has been obtained. Diaphragms spread the heat equally among the stacks of upright tubes.

The half section at the right shows the tube connections with the shell.

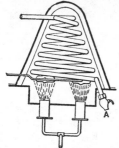

84. FLASH TYPE STEAM GENERATOR. The water is fed at the bottom of the coil at A. Gasoline is vaporized in the small cast iron retorts B, placed beneath the steam coil. A cheap and safe type for amateur and automobile use. The generation of steam is controlled by the pump action in this class of boilers.

85. Plan of retorts and connections from the feed pump and to the burners beneath the retorts.

86. NOVEL MOTOR. In this motor bulbs are arranged diametrically opposite each other, in pairs, each pair being connected by a

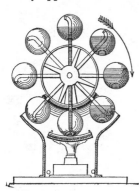

tube. The motor thus formed of the series of bulbs, the tubular arms and the shaft supporting them, is operated by the heat of a small lamp. Each pair of bulbs contains enough water to fill one of them. The wheel thus formed revolves over a deflector which is heated by means of the lamp. The bulbs are exhausted of air, so that pressure sufficient to force the water from the hot bulb to the cooler one above quickly generates from water under a vacuum by its low boiling point.

87. SOLAR CALORIC ENGINE. Ericsson system. This engine ran at 420 revolutions per minute in clear sunlight. It was con-

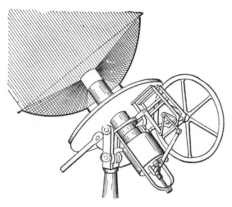

structed on the same design as the ordinary hot-air engine and ran under the same conditions.

It is calculated that the heat radiated by the sun during nine hours per day, for all the latitudes comprised between the equator and the 45th parallel, corresponds per minute and per square foot of normal surface to the direction of the rays to 3·5 thermo units of 772 foot pounds. Hence, a surface of 100 square feet would give a power of 270,000 foot pounds, or from 8 to 9 horse-power.

88. MOUCHOT'S SOLAR BOILER. A is a glass bell, B is a boiler with a double envelope, D is a steam pipe, E is a feed pipe, F

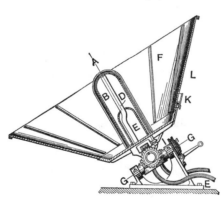

is a conical silvered mirror; G G is a spindle around which a motion is given to the machine from east to west by the gearing regulating the inclination of the apparatus on the spindle G G, according to the seasons; I is a safety valve; K is a pressure gauge, and L is a water gauge.

Diameter of top 9 feet; 45 square feet of silvered glass surface. Boiler of blackened copper 31 inches high, 11 inches diameter. Thin glass cover 2 inches larger than boiler.

Pressure generated 75 lbs. per square inch; 11 lbs. of water evaporated per hour. Used for driving a pump. A sun motor of this type is now in operation in Southern California, pumping water for irrigation. Reflector 33½ feet diameter, 10 horse-power.

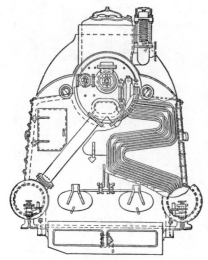

89. MARINE WATER TUBE BOILER. A light and powerful boiler invented by Du Temple in France and used on an English torpedo gunboat. Patented 1876.

This boiler has all the essential qualities of the later water tube boilers.

Ample water circulation is provided for by the back connections, one of which is shown in the cut.

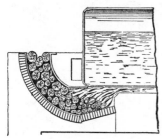

90. DOWN DRAUGHT WOOD-BURNING FURNACE. The curved chute facilitates the self feeding of the wood to the grate. Width of chute suitable for cord wood. Fire trimmed from a side door. St. Clair type, which is also adapted to the burning of bituminous coal.

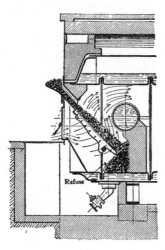

91. GRAVITY FEED FURNACE. For an internal fire-box boiler. For bituminous coal the inclination of the grate made to suit the sliding properties of the coal. The feed hopper extends clear across the grate width. The coal feeds down by rate of combustion, which in turn is regulated by the amount of draft admitted. The new coal is heated by the burning fuel before it properly "catches," and thus a preliminary evolution of gas is effected, which lessens very perceptibly the amount of visible smoke given off by the furnace.

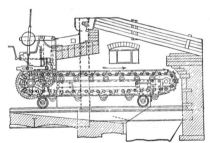

92. TRAVELING LINK GRATE. The link-bar grate is fed forward by a geared drum carrying the coal fed from a hopper and coked under the fore arch of the furnace. Motion of grate and amount of coal regulated by speed of gear and opening of the hopper, sliding door, and grate guard.

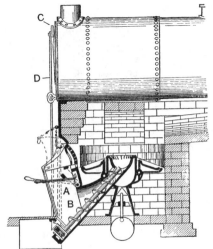

93. UNDER FEED FURNACE. A circular grate with a central recess to which the coal is lifted from the magazine by a spiral carrier. The coal is pushed up through the central funnel and falls over on to the grate, which is circular.

A, magazine or hopper.

B, feed screw.

94. DOWN DRAUGHT FURNACE in an internal fired boiler. Eastwood type.

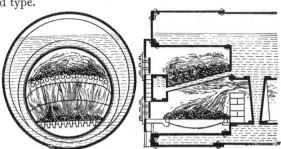

A water tube grate with tubes between the furnace head and a cross-head between the doors of the upper and lower furnace for obtaining a perfect circulation in the grate.

95. Longitudinal section of furnace.

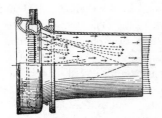

96. ANNULAR STEAM BLOW-ER. For boiler and other furnaces. An annular cast-iron chamber perforated for steam jets at an angle that projects the jets in a converging cone and draws in the air with a force corresponding with the pressure of the steam.

97. STEAM BLOWER. Eynon-Korting type. A double nozzle air injector and double cone tube for boiler and other furnaces. The

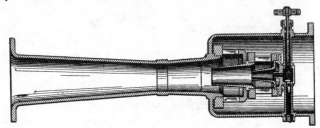

needle valve regulates the flow of steam from the central jet which is re-enforced by the incoming air around the two nozzles.

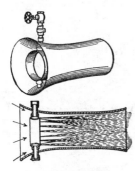

98. ARGAND STEAM BLOWER for furnaces. A perforated annular nozzle in-closed in a shell with curved sides and steam connections.

It furnishes a large volume of air with a small amount of steam. The air and steam are thoroughly mixed in the shell of the blow-er before the blast is delivered into the ash pit. It makes very little noise in operation.

99. Section, showing ring and jets.

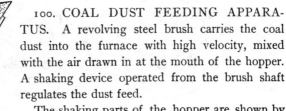

100. COAL DUST FEEDING APPARA-TUS. A revolving steel brush carries the coal dust into the furnace with high velocity, mixed with the air drawn in at the mouth of the hopper. A shaking device operated from the brush shaft regulates the dust feed.

The shaking parts of the hopper are shown by the dotted lines.

101. COAL DUST BURNER. The vertical tube, *a*, for introducing the air serves as a pivot to the system, which comprises : 1, the movable jacket, *b*, to which are bolted the primary air conduit, *c*, and the secondary air conduit, *d;* 2, the hopper, *e*, supported by the chest, *f;* the inlet pipe, *g*, and, finally, the conical chamber, *h*. The chest, *f*, is supported by the conduit, *d*, but is not in communication with it.

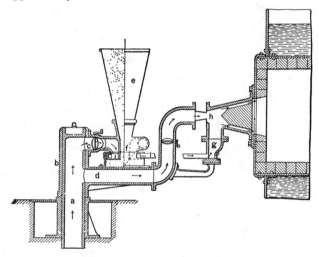

The fixed pipe, *a*, is provided at the side with two apertures in front of which the two conduits, *c* and *d*, coincide when the apparatus is in operation. At a stoppage, a rotary motion is given the system and the apertures of the pipe, *d*, are closed by the sides of the movable jacket. The conduit, *c*, is divided into two branches which discharge the dust into the inlet, *g*.

102. FUEL OIL BURNER. Plan and section of the crude oil burner used on the Southern California Railroad. This burner has but two chambers, oil and steam; the air enters the furnace through graduated openings around the burner.

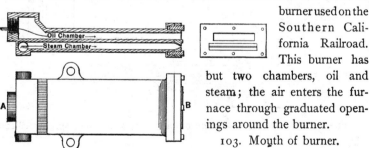

103. Mouth of burner.

104. Plan of burner.

105. BURNER FOR AUTO-BOIL-ER. The disk chamber is stamped out of plate iron. The interior air tubes are screwed into the back head and furnish air to complete combustion. The outer tubes are screwed into the top plate and furnish an annular stream of vapor gas from the chamber below.

106. AUTOMOBILE BOILER. Showing the arrangement of burner and vaporizing coil above the burner tubes.

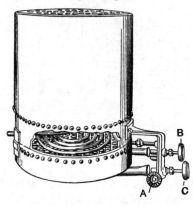

The gasoline is fed to and vaporized in the spiral coil. The vapor injected into the burner chamber by the valve B carries air with it. C and A are oil and air atomizing valves for starting the burner.

The gasoline tank should have an air pressure of 30 pounds. The atomizing valve A is connected with the air chamber of the gasoline tank.

107. FUEL OIL BURNER. Plan and sections of the crude oil burners used on the locomotives of the Southern Pacific Railroad.

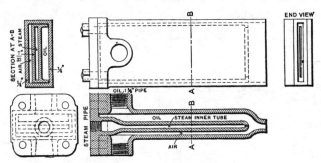

The steam, oil, and air chambers are very wide and spread a broad flame.

Nos. 108, 109, 110, 111, show the burner in detail.

112. LIQUID FUEL BURNER. Urquhart Locomotive type.

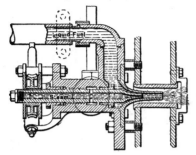

The air nozzles are fixed, steam nozzle is movable by a screw and worm gear and regulates the oil flow. Air enters between the front of the boiler and a plate held off by studs. A stay tube through the water space makes an entrance of the flame to the furnace.

113. OIL FUEL FURNACE, for heating and setting tires. The combined oil and air enter the hood through a peculiarly shaped

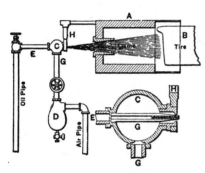

expansion nozzle, which effectually combines them and spreads them outward at the same time. The air pipe connects with a chamber D, which has a cock for draining off water. A valve is provided between D and C for controlling the pressure. The oil pipe connects to the back of C, and is carried through C by a pipe G which terminates with a nozzle having four holes. The air escapes through an annular orifice, surrounding the end of the nozzle, and carries the oil in the form of finely divided spray through the expansion tube into the hood A.

114. OIL FUEL BURNER. An English type of approved

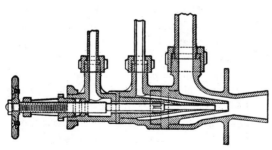

design in which the steam issues in an annulus within an annulus of oil and air at the throat of a double cone. Designed for the most economical combustion of oil.

115. FUEL OIL BURNER. An air burner. Brown type. The air issues in an annular converging cone *d*, meeting the small jets

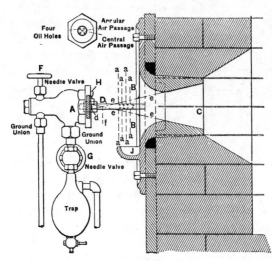

of oil from the central apertures. The central tube D also ejects through the small holes *a, a, a, a,* fine jets of heated oil; all forming the diverging cone *e, e, e, e,* of atomized fuel. C is a flared opening of fire brick. J, a cup for igniting. Air pressure, 25 to 30 pounds.

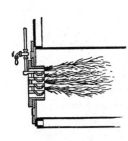

116. LIQUID FUEL BURNER. Cup grate type, operated by natural draught of the chimney. The heat of the furnace heats the inner edge of the cups or troughs, vaporizes the oil, which mixes with the air in draught. The oil is fed to the upper trough and overflows to the next trough, and so on for as many troughs as required. A sliding cover and air-stopper regulates the fire. See 117 for details.

117. PETROLEUM FIRE GRATE. Nobel type. This grate consists of a series of superposed troughs, *a, a', a'',* containing the

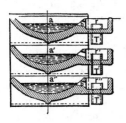

liquid fuel, the discharge of which is regulated by small basins, *r, r', r'',* that communicate with the troughs at *b, b'.* Through the basins pass discharge pipes, T, T', T'', open at each extremity. All the parts of the grate are of iron cast in a piece, and comprising no movable joint. The oil enters the top trough and overflows through the tube, T, to the next, and so on through the last overflow to a receiving tank. In this manner all the troughs are kept at constant level.

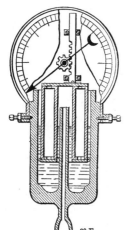

90 E

118. CHIMNEY DRAUGHT IN-DICATOR. The inverted float in a cup of water is connected to the dial hand by a rack and pinion. The flue or chimney is connected to the bottom of the central tube by pipe or hose. The draught of the chimney relieves the float of pressure equal to the static height of the water due to the partial vacuum, which becomes a constant record on the dial.

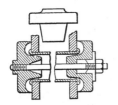

119. PLUG FOR LEAKY BOILER TUBES. The bolt is screwed into one of the plugs tight; the other plug has a gland and stuffing box for perfect closure around the rod by rubber or asbestos packing.

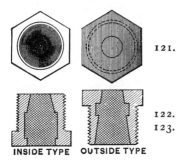

INSIDE TYPE   OUTSIDE TYPE

120. SAFETY PLUGS FOR BOILERS. Lunkenheimer type to **121.** meet the requirement of the United States Inspection Service, which prohibits alloys as not being reliable, and requires that all plugs shall be **122.** filled with pure Banca tin, which **123.** melts at a temperature of 446° Fah.

INLET

124. SIMPLE FLOAT STEAM TRAP. Eureka type. A tight copper ball with valve and guide stem at the bottom and guide stem in the inlet pipe.

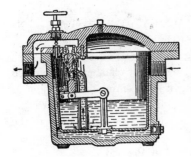

125. AUTOMATIC STEAM TRAP. Lawler type. The open float by its overflow and filling, sinks, and opens the discharge valve by its lever connection. When the float is emptied by the steam pressure, the float rises and closes the valve.

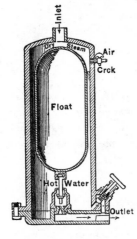

126. FLOAT STEAM TRAP. One of the several types of steam traps with sealed floats, this having a direct attachment of the float and valve which is designed to give a small amount of motion to the float for operating the valve.

The valve has a cage guide and its stem is loosely socketed to the float, so that any side motion of the float does not unseat the valve.

127. DIFFERENTIAL EXPANSION STEAM TRAP. The opening and closing of the valve for discharging the water of condensa-

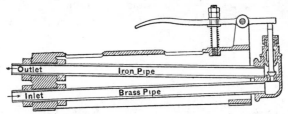

tion is effected by the differential expansion and contraction of the brass and iron tube, being 3 to 2. The setting of the valve is controlled by the adjusting lever, so that the water is discharged, and when steam enters the brass tube it expands by the additional heat and closes the valve by lifting the seat.

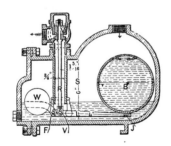

128. BALANCED STEAM TRAP. The float B is always full of water and not liable to collapse. It is balanced by the counterweight W, at the other end of the valve lever, so that the float B opens the valve by the differential flotation of the weight and float, when the chamber, S, fills with water.

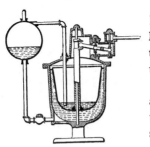

129. RETURN TRAP. Blessing type. For raising the water of condensation to a higher level than the water line of a boiler, to be returned to the boiler by gravity under equalized pressure.

The movable bucket operates a lever and the equalizing valves for discharging the water to a receiver by the boiler pressure and from the receiver to the boiler by gravity.

130. AUTOMATIC BOILER FEEDER. Feeds water to a boiler on the same principle as the pulsometer.

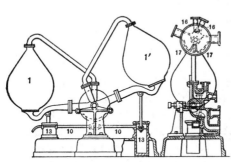

The feeder is placed about 4 feet above the water line of the boiler with water flowing to it by gravity or pressure. The weight of the water alternately filling the chambers, carries them down and opens a steam port in the axial valve to the boiler pressure, when the water flows into the boiler by gravity. At the same time the upper chamber is being filled by the condensation of its steam. It is made self-acting by having the steam pipe connect to the boiler at the high water line, at which point steam can not enter the chamber and the action stops. The dashpots regulate the motion of the feeder.

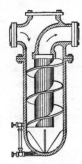

131. CENTRIFUGAL STEAM SEPARATOR.

The centrifugal force produced by the steam whirling around the spiral partitions causes the entrained water to be thrown against the outer shell to drip to the bottom of the case.

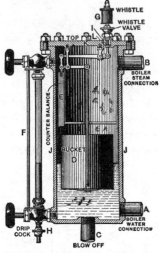

132. LOW WATER ALARM.

Bundy type. For steam boilers. An overbalanced submerged bucket with lever attachment to the whistle valve. Low water uncovers the bucket, when its unbalanced weight opens the whistle valve.

The parts are detailed in the cut.

133. SIMPLE BOILER FEED DEVICE.

A is a vent cock to discharge air or steam from the feed tank ; C is a cock to let steam

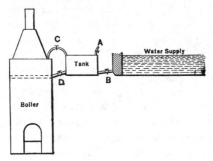

from the boiler to the tank ; D is a cock to let water from the tank to the boiler. The bottom of the tank should be above the highest water level to be carried in the boiler. The lowest level of the water supply must be higher than the bottom of the tank. To feed the boiler, first close C and D and open A and B. Water will then run into the tank. Then close A and B and open C and D, and the water in the tank will run into the boiler.

134. FEED WATER HEATER AND PURIFIER. Anderson type. This apparatus consists of a vertical cylinder containing a

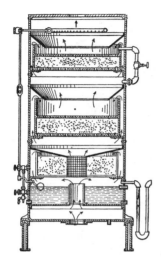

number of compartments filled with filtering material. The exhaust steam enters at the bottom and flows into the first compartment through a short pipe, thence through the annular opening surrounding the second compartment, into the latter, thence through another annular opening into the next compartment and so on to the top of the cylinder or casing.

After passing through the annular openings, the steam comes in contact with baffle plates, which direct the steam through the falling water, thus condensing a large part of the steam. The water enters at the top through perforations in a ring-pipe, the water falling upon a baffle plate, which delivers it into the upper filtering compartment. From the latter compartment the water falls in drops through the current of steam into the second filtering chamber or bed and so on to the storage reservoir at the bottom. A ball float is connected with the water-regulating valve at the top and maintains a constant water level in the storage reservoir. A sealed overflow pipe prevents the water in the reservoir from overflowing into the exhaust pipe. The feed pump takes the hot water from the reservoir at a point near the bottom, thus avoiding any oil that might be present at the surface.

135. SURFACE CONDENSER. Double tube type in which the shell has both tube heads at one end ; the cooling water flowing

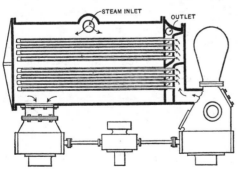

through the smaller concentric tubes and returning through the annular space between the tubes. There being but one joint for each tube, the troubles arising from expansion and contraction of the tubes are avoided.

136. NOVEL SURFACE CONDENSER. A cylinder filled with small tubes as shown in the section, **137.** A spray jet of water

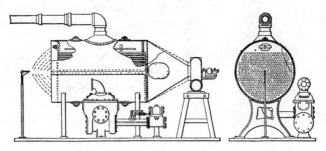

thrown against the tubes at one end, and with a large volume of air, is drawn through the tubes by a suction blower at the end of the conical chamber. The water is vaporized and with the air takes up the heat of the exhaust steam and is discharged in a vapor by the blower. The pump keeps up the vacuum in the exhaust chamber and returns the water of condensation to the boiler. Claimed to use but one pound of water for each pound of steam condensed.

137. Section of condenser and tubes.

138. EVAPORATOR for obtaining fresh water from salt water. The chamber is kept supplied half full of salt water and kept below

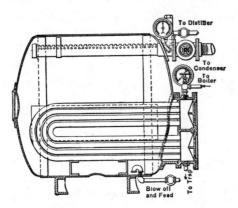

saturation by blowing off. The vapor is drawn off through the perforated pipe at the top through a condenser by the vacuum pump. The boiling temperature of the salt water of the ocean is about 153° Fah. at a 26-inch vacuum. The condensed steam from the coils is saved and filtered or again fed to the boilers on ship board. Enough vapor for ship use is conveyed to an aërator and cooler.

SECTION V.

STEAM POWER APPLIANCES.

71

Section V.

STEAM POWER APPLIANCES.

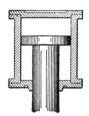

139. TYPES OF COMPOUND ENGINES. Single cylinder or trunk engine in which the high and low pressure areas are adjusted by the size of the piston rod or trunk, which is inclosed by a stuffing box. The connecting rod is jointed within the trunk.

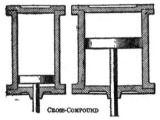

Cross-Compound

140. TYPES OF COMPOUND ENGINES. Cross compound with cranks at 120° on ends of shaft. Fly wheel in center. In this type there is no dead center.

141. Low-pressure cylinder, showing relative positions of piston.

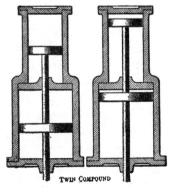

Twin Compound

142. TYPES OF COMPOUND ENGINES. Twin compound, close connected tandem with high-pressure cylinders forward, cranks at 120° on ends of shaft, fly wheel in center.

143. The relative positions of the pistons are shown in the two figures at alternate strokes.

73

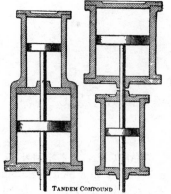

TANDEM COMPOUND

144. TYPES OF COMPOUND ENGINES. Tandem compound types in which the forward cylinder is high pressure with close connection to the low-pressure cylinder with only a metallic sleeve as a stuffing box.

145. In the reverse type the cylinders are independent and separated a short distance with regular stuffing boxes.

146. TRIPLE EXPANSION ENGINE, with double tandem high-pressure cylinder. Edwards patent. The object of the design is

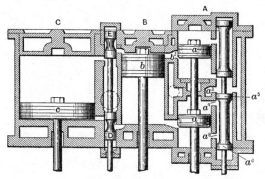

to produce an arrangement of cylinders, steam valves, and ports whereby the back pressure of the intermediate cylinder will not act as an opposing force on the high-pressure piston, and will also furnish full pressure of steam in the intermediate without increasing back pressure in the high. Steam enters the chamber a^3, passes through an opening between the two piston valves, which opened to the upper piston, a, when it passed the bottom center. The cut shows it in the act of closing.

When working as a triple expansion the valve closes when the piston reaches the point b^2, which allows the steam to enter cylinder B above piston b at full pressure, but the crank to cylinder A is on the quarter where it moves at its highest speed, while the piston B moves down. It will also be seen that lower piston A reaches the top of its cylinder. At the same time, but instead of being in a position to exhaust as in the upper one, it will be in the position to receive through lower port q^9, valve a^5 having moved down far enough to open. The piston A starts on the return stroke, which requires no explanation.

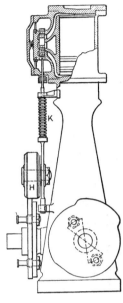

147. HIGH SPEED VERTICAL EN-GINE. Rhodes type. Valve of the gridiron balanced type, with valve gear of a novel design giving a quick and full movement of the valve through a rawhide disk, H, rolling on an irregular cam on the shaft and made adjustable for time of cut off between the limit of $^3/_8$ and $^5/_8$. The roller is kept in contact with the cam by the spring K.

The cam is shown on the engine frame.

148. COMPOUND STEAM OR AIR ENGINE. Watson type. The steam or air is admitted from the bottom and between the

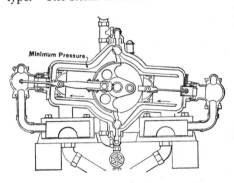

pistons, driving the pistons outward to the position as shown at the left end of cylinder. When the pistons are in this position, the cam valve opens the top port and closes the feed port, thus allowing the steam or air to collect between pistons and pass to the opposite side of the same, driving them inward to the position as shown in the right end of cut. As soon as the inward stroke is completed, the extension stem on pistons opens the end or exhaust valves, which remain open until the pistons complete their outward stroke and cycle of revolution. To reverse the motor, steam or air is taken from the top instead of the bottom. The motors are so arranged that steam or air can be used once and exhaust, or it may be used twice.

149. TRIPLE EXPANSION MARINE ENGINE.   Type of steamer Minnesota.   Proportion of cylinders, 1, 5, 15 in area; stroke,

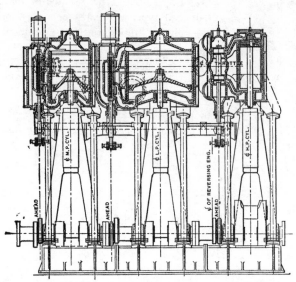

48 inches; crank positions, 120°; high-pressure cylinder, 23 inches diameter; intermediate cylinder, 51 inches diameter; low-pressure cylinder, 89 inches diameter.

150. COMPOUND CORLISS ENGINE.   Atlas type, with direct connected exhaust valves for both high- and low-pressure

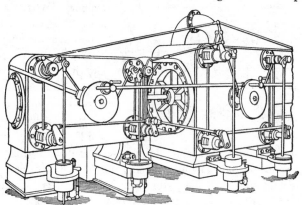

cylinder.   The releasing gear of this type of engine is unique in that inertia, centrifugal force, and the force of gravity are used to operate the grab hook.

151. COMPOUND CORLISS ENGINE. Showing Corliss

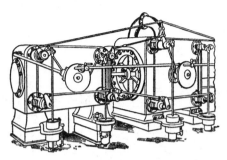

valves on low-pressure cylinder, direct connected wrist plates, one for each cylinder to operate the steam valves while the exhaust valves are all connected by link rods, which are in turn directly connected to the eccentric rod. Valve gear of both cylinders have trip hooks and dashpots.

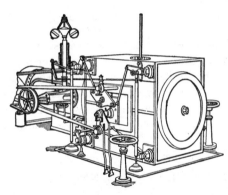

152. CORLISS EN-GINE. C. & G. Cooper type with double wrist plate. The governor controls the action of the steam valves by adjustable trip hooks and dashpots.

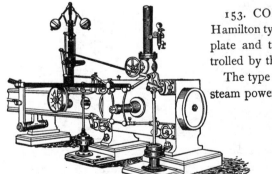

153. CORLISS ENGINE. Hamilton type, with single wrist plate and trip-valve gear controlled by the governor.

The type of most economical steam power.

154. CONVERTIBLE COMPOUND ENGINE.   Flinn type for automobiles.   Steam enters at the center of the high-pressure steam

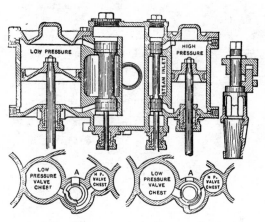

valve, and when the intercepting valve is in the position shown in the left cross section it can pass from the high-pressure chest directly to the low-pressure chest, allowing both cylinders to run with high-pressure steam as simple engines, the high pressure exhausting at A into the main exhaust chest.   This gives the machine great starting or climbing power.   When less power and more economy is wanted, the intercepting valve is turned to the position shown in the right section, closing the free exhaust from the high-pressure cylinder and the live pressure connection to the low-pressure steam chest and compelling the exhaust of the high-pressure cylinder to enter the receiver and flow to the low-pressure valve.

155. A vertical section of the intercepting valves and ports.

156. NOVEL THREE-CYLINDER ENGINE.   The novel features are in the manner in which the piston valves are operated and

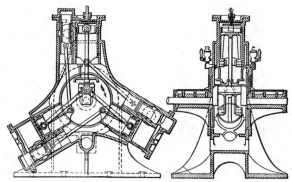

in supplementary exhaust ports.   The piston valves are connected to and operated by the following piston.   The exhaust is discharged into the main trunk of the engine through the hollow spool valves and from the ports opened by the trunk pistons into the jacketed recesses, making its final exit at the bottom of the casing. A compact high speed engine of English design.

157. Vertical section on line of shaft.

158. REVOLVING ENGINE. Kipp type. In Kipp's re-

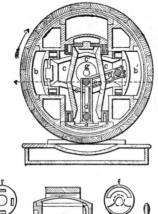

volving engines the exterior cylinder, to which a belt may be directly applied, it being surrounded by·a lagging for that purpose, is caused to rotate by the reciprocation of two pistons with duplicate heads in cylinders whose axes are at right angles to each other. The piston heads *a a'* are connected, as are also *b b'*, by the pieces *c c c' c'*. Yokes *d d'* connect these with a crank *e* on the main shaft of the trunk. Steam is admitted through the valve *f* to the central space *g*, which serves as a steam chest. The arrangement of the ports is shown at *i*. The drum is mounted on trunnions, through one of which the steam enters, the other serving to exhaust through one of the hollow pillars *k* into the feed-water heater *l;* an eccentric on the main shaft also operates the feed-water pump.

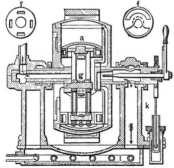

159. Section through axis of rotation.

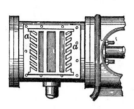

160. FRICTION RELIEF IN D VALVES. This novel method of relieving the friction of slide valves consists in cutting diagonal grooves in the outer bearings of the port face of the steam chest, as shown at *a, a*. This relieves the pressure of the valve and facilitates lubrication.

161. NOVEL TRIPLE COMPOUND MARINE ENGINE.
The novel features are the three-part eccentric oscillating upon the

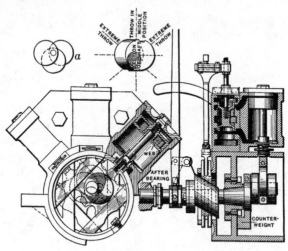

crank pin and upon each of which a strap fixed to the piston rod of
each cylinder slides in ways parallel with each piston rod.   The throw
of eccentrics and crank are each equal to one-half the piston stroke.
The eccentrics are at 90° and 180°, as shown at *a*.   The three piston
valves are directly connected by rods to thin straps on an angularly
mounted cylinder that slides on the shaft by the hand lever for forward, stop, or reverse motion.

Piston valves are used, taking the steam in the middle and exhausting at the ends.   The steam passes from the first valve, through the
triangular space between the cylinders, to the next valve chest.

162. Vertical section through intermediate cylinder.

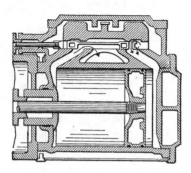

163. TYPES OF SLIDE
VALVES. Slide valve of the
Ames engine. The valve is finished on both sides and rides under
a partly balanced pressure plate.

164. BALANCED PISTON VALVE. The segmental packing, E, is held close to the cylinder wall by the pressure of the steam which

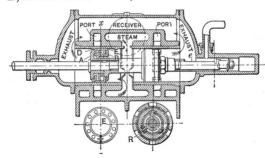

enters through holes in the flange, E, shown by the circles below the cylinder. e, live steam connection to receiver of low-pressure cylinders, a supplementary valve operated by the stem of the piston valve. Used on compound locomotive, Italian railway.

165. TANDEM COMPOUND LOCOMOTIVE CYLINDERS. Balanced valves. Type of Pittsburgh Locomotive Works.

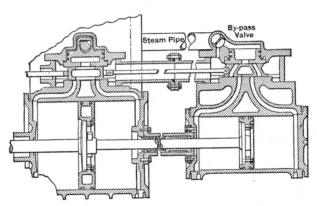

The cylinders are separated somewhat and have a sleeve between the heads, which is bolted to the front head of low-pressure cylinder. At the front it is held by a flange which makes a joint around it. This allows easy inspection and repair of low-pressure piston, as the sleeve in question slides into the high-pressure cylinder and both pistons can be moved forward together and out of the cylinders.

The valves are connected by a rod passing through a pipe between the steam chests of the high-pressure and low-pressure cylinders. The high-pressure valves receive steam through the balance plate, which is fitted into the chest cover. The steam goes through the ports in the valve to the passages in the cylinders.

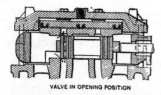

VALVE IN OPENING POSITION

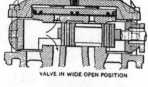

VALVE IN WIDE OPEN POSITION

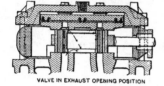

VALVE IN EXHAUST OPENING POSITION

166. BALANCED VALVE for steam engine. Wilson type. Pressures are equalized by steam pressure under the riding plate. Valve has double admission and double exhaust ports. The three sections show the positions of the valve when opening, wide open, and exhaust opening position.

167. Wide open position, taking steam under the balance plate.

168. Position of valve at exhaust opening of both cylinder and balance plate.

169. NOVEL PISTON VALVE for a steam engine. A side elevation, partly in section, of the valve and its casing, and a longitu-

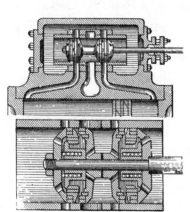

dinal sectional elevation. The valve consists principally of a relief valve held seated by a spring, but exposed at the opposite side to the pressure of the steam, so that in case of excessive pressure, suction, or vacuum, the engine being in motion and steam shut off, the valve will lift, and steam or hot vapor and gases will enter and destroy compression and vacuum, by way of the apertures under the valve leading to the exhaust, as well as by the opening directly into the steam pipe through the piston valve, thereby giving free openings from the steam pipe direct to the atmosphere through the exhaust pipes. The valve casing is formed with steam ports, and the valve is composed of two similar heads or pistons, each formed of a circular plate with a rim parallel with the piston rod and a vertical flange, there being openings through the plate and through the rim. The relief valve is fitted within the rim and held to its two seats by a coiled spring held in contact with the valve by a circular plate on the piston rod, which also holds the packing rings and an outer ring firmly against the flange of the body of the valve. One seat of the relief valve covers holes leading to the exhaust, and the other seat, upon its outer rim, covers the passage leading into the steam pipe and chest, the unseating of the valve opening all connecting passages through the piston head, including passages from the throttle to the escape pipe or atmosphere, simultaneously.

170. AUTOMATIC VALVE MOTION. For a steam pump. The striking of the supplementary valves in the cylinder head by the

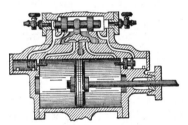

piston releases the pressure on that end of the valve bobbin, when it is thrown over, carrying the valve with it. The small cylinders in each end of the main cylinder have each a live steam port and an exhaust, and within them pistons work freely as independent valves, each having a stem normally projecting within the main cylinder. These valves are operated in one direction by the main piston coming in contact with their stems, and are moved by the pressure of steam on their backs in an opposite direction. It is applicable to direct-acting pumps, and also to direct-acting engines for other than pumping purposes.

171. TYPES OF SLIDE VALVES. Slide valve of the Chandler & Taylor tandem compound engine. The valve is of the gridiron type,

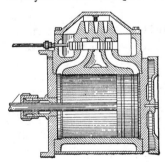

and is double-ported for both steam and exhaust, making it possible to admit large amounts of steam into each end of the cylinder quickly and with a very short valve travel. The valve, being light and perfectly balanced by means of the pressure plate on its top side, is therefore very easily acted upon by the governor.

172. TYPES OF SLIDE VALVES. Slide valve of the Brownell engine. The valve is of the box type, double-ported for both steam 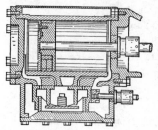 and exhaust, and practically perfectly balanced. The steam pressure is removed from the back of the valve by means of a balance ring which bears against the steam-chest cover. A coil spring serves to keep the ring against the chest cover, thus taking up the wear automatically and preventing the ring from leaving the seat and causing annoyance by rattling.

173. CONCENTRIC VALVES, CORLISS TYPE. This valve, although essentially of the " Corliss " class, differs from the ordinary in 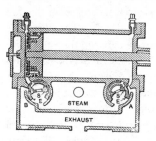 that the steam valve is inclosed in the exhaust valve, making practically only two valves, which, however, perform the functions of four perfectly. A cross section through the cylinder and valves is presented, where E is the exhaust valve and S the steam valve. The steam valve, of the double-ported balanced type, is held within the exhaust valve E, but is not set exactly in the center of the latter valve, so that it is held in position by steam pressure. The usual vacuum dash-pots are replaced by spring dashpots—that is, the tension of springs is relied upon to close the valves, while the air cushioned in the dashpot cylinder prevents the shock which would be inevitable were it not used.

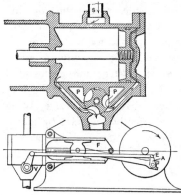

174. OSCILLATING STEAM AND EXHAUST VALVE, for hoisting engines. The valve is operated by a direct rod from crank-pin arm to the valve arm. S, steam pipe with passage around the cylinder to the steam chest, P, P. A good design to keep the cylinder clear of water.

175. Shows connection from crank-pin arm to valve arm.

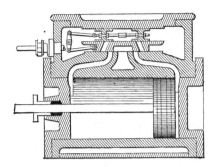

176. RIDING CUT-OFF VALVE. From single eccentric. The main valve is moved by the direct connected valve rod. The riding valve is moved by a short lever and links pivoted to the two valves.

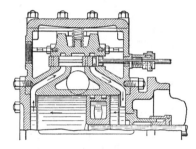

177. TYPES OF SLIDE VALVES. Slide valve of the Bayley engine. A flat valve riding under a balanced pressure plate. Pressure plate is held in place by stays against the steam chest.

178. PARSON'S STEAM TURBINE. Steam is admitted at the governor valve and arrives at the chamber, A, at the small end of the

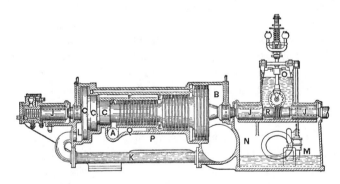

revolving part of the turbine. The steam passes along to the right through the turbine blades, passing through a series of fixed blades which deflect it in one direction, thence striking the moving blades of the turbine which deflect it in the opposite direction, and so on. In this way the current of steam impinging upon the moving blades drives them around. The areas of the passages increase, progressing in volume corresponding with the expansion of the steam. On the left of the steam inlet are revolving balance pistons, CCC, one corresponding to each of the cylinders in the turbine. The entering steam at A presses equally against the revolving part of the turbine and against the first balancing piston. When it arrives at the passage, E, it presses against the next larger section of the revolving part of the turbine and also against the next largest balancing piston, connection between the two being secured by the passage, F. Similarly, the passage, G, permits the balancing of the largest section of the turbine. By a proper arrangement of these balancing pistons there is no end-thrust upon the turbine shaft at any load or steam pressure. The thrust bearing at H, on the extreme left, is to take care of accidental thrusts that may arise and also to retain the alignment of the shaft accurately so as to secure the correct adjustment of the balance pistons.

Since these balance pistons never come in mechanical contact with the cylinder in which they turn, there is no friction. The thrust bearing is made of ample size and is subject to forced lubrication.

The pipe, K, connects the chamber back of the balance pistons with the exhaust outlet, so as to insure the pressure being equal at the two ends of the turbine.

The bearings, JJ, are peculiar in construction. Each consists of a gun-metal sleeve prevented from turning by a loose-fitting dowel pin. Outside of this are three cylindrical tubes having a small clearance between them. These small clearances fill up with oil and permit a slight vibration of the inner shell, while at the same time restraining it from too great a movement. The shaft therefore actually revolves about its axis of gravity instead of its geometrical axis, as would be the case with the bearings of the usual rigid construction. In case the shaft is a little out of balance the journal thus permits it to run slightly eccentric. The form of the rotating and stationary blades are much like those of the Curtis type, which are detailed in the following cuts.

The economy of the steam turbine has been greatly advanced by improvements since its advent, so that it is now nearly in line with the best quadruple expansion engines, and with it the highest speeds in navigation have been obtained.

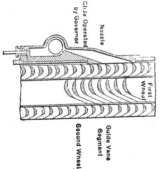

179. STEAM TURBINE. Curtis type, showing the arrangement of the steam passages in the moving and stationary blades in a three-disk engine.

Claims are made that this type of turbine with vacuum exhaust uses but 12 pounds of steam per horse-power. The diverging nozzle is made of variable area by a slide valve and governor.

180. A segment of one of the disks shown on a larger scale. The blades of the segments are cut in a milling machine of special design, and are bolted to the rim of the disk. A band incloses the outer end of the blades to prevent undue leakage between the disk and shell.

181. STEAM TURBINE. Multinozzle type. Showing position

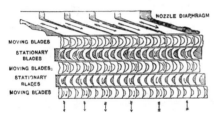

of blades in reverse curves on the moving and stationary disks. The multinozzle may extend all round the disk, as in the first stationary disk of each section of the Parson's turbine.

182. STEAM TURBINE. De Laval type. Vertical section showing form of buckets and nozzles. Steam impinges against the outer edge of the buckets and exhausts at the sides.

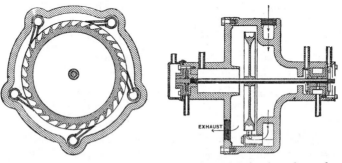

183. Plan showing spring shaft, bearings, lubricating channels and

steam ducts. Runs by the impact of steam from five nozzles against the outer edge of the buckets of the wheel. The long shaft is to take up the unbalanced vibration of the disk.

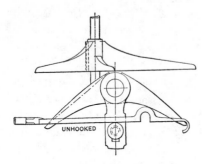

184. THE STEVENS VALVE GEAR. Showing the double toe and wipers with the eccentric rod unhooked. Type used on the Hudson River steamers. First used in 1840. A standard type for marine walking-beam engines.

185. VALVE GEAR. A wrist plate journaled on a pin carried by a standard or post on the engine frame. The wrist plate is rotated

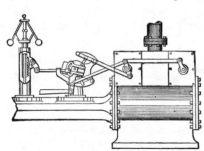

by the eccentric, the motion being communicated to the several valves by suitable rods which are connected to two horizontal links pivoted to the wrist plate ; these links being held outward in their proper position by two vertical links the inner ends of which are pivoted to the slotted bar near the wrist-plate hub. The rod from the governor passes through a guide carried by the post and is connected to the slotted bar by means of a sliding block working in a concentric slot as shown, which permits the bar to oscillate with the wrist plate without interfering with the governor rod. It will be seen that when the slotted bar operated by the governor occupies the position shown, the valves have full travel, but when this bar is drawn toward the governor the ends of the horizontal links to which the valve rods are attached will be drawn toward the hub of the wrist plate by the short vertical links, thus reducing the radius of the valve-rod connections which shortens the stroke of the valves and consequently changes the point of cut-off in the cylinder.

186. CORLISS VALVE GEAR and release mechanism, standard type. A, valve stem.

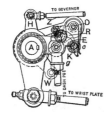

A bell crank operated by a connecting rod from the wrist plate, lifts the grab hook, E, and the valve arm. An adjustable roller at R, releases the valve arm, which is pivoted to the dashpot for regulating its fall. The release roller is operated by the bell crank H, and rod Z, from the governor.

187. CORLISS VALVE GEAR and release mechanism. The grab hook consists of a block, C, sliding in a grooved slot in the bell-

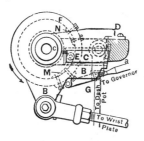

crank lever B B, and normally forced outward by a spring. The block C carries a pin, E, on the rear side, which is held in contact with a cam ring, F, having two knock-off dies, M and N, on its inside surface. As the bell crank moves in the direction of the arrow from the position shown, the roller on the pin E strikes the cam die N, and is forced rapidly inward, releasing the drop lever *a*. If from any cause the dashpot should fail to act, the projection on the bell-crank lever would engage with the drop lever and close the valve.

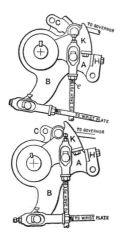

188. CORLISS VALVE GEAR. In this design, B is the bell crank, which carries the hook H mounted on a short shaft, on the other end of which is the trip lever (not shown), which engages with the knock-off cam C, operated by the governor rod. K is the drop lever with dashpot connection. The cam lever C, controlled by the governor, limits the time of release of the hook H.

189. Shows the position of the parts at the moment of release.

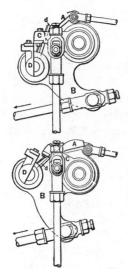

190. CORLISS VALVE GEAR. This design consists principally of a curved bell crank, B, carrying the grab hook D mounted on a short shaft having an arm at the other end. The trip lever *d* rides on the knock-off cam A, the position of which is controlled by the governor, as usual. When the bell crank reaches the position shown in the upper sketch, the trip lever is thrown outward, releasing the drop lever, the point of release being governed by the position of the knock-off cam.

191. Shows the position of the parts at the moment of release.

192. CORLISS VALVE GEAR. A is a bell-crank lever mounted loosely on the valve stem or on a projection of the bonnet, and carries

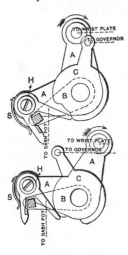

the grab hook H at one end and is connected to the wrist plate by an adjustable connecting rod, from which it receives its motion. The hook H is normally pressed inward by the spring S so that the longer arm of the hook is always held firmly against the knock-off cam C, which is placed next to the bell crank and is connected to the governor by a reach rod. The drop lever B is keyed to the valve stem and connected to the dashpot by a rod ; it carries a steel block or die which engages with the block or die on the grab hook H. As the bell crank A moves in the direction of the arrow, the hook is engaged with the die on the drop lever B, and as their relative positions remain constant, they having a common center of rotation, the end of B is raised, opening the valve, which remains open until the bell crank has advanced so far that the longer arm of the hook H is pressed outward by the projection on the knock-off cam C, when the drop lever B is quickly brought to its original position and the valve is thereby closed.

193. Shows the position of the parts at the moment of release.

194. CORLISS VALVE GEAR. Allis-Chalmers type. Starting from the lowest position (not shown), the hook H, which is forced

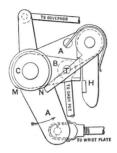

inward by the spring, engages with the drop lever B, and as the bell-crank lever, A, A, moves in the direction indicated by the arrow, the lever B is carried around to the position shown, opening the valve. When this position is reached, the trip lever T comes in contact with the projection N of the cam C, forcing it, and consequently the grab hook H, outward, and releasing the drop lever B, which is rapidly brought to its original position by the action of the dashpot.

195. DASHPOT FOR CORLISS ENGINE. As the plunger, P, is drawn upward by the valve gear, air is drawn into the plunger

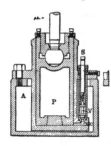

cylinder from the annular chamber, A, through the check valve C. The air is not sufficient, however, to prevent the formation of a partial vacuum which draws the plunger quickly downward when the valve spindle is released. As the plunger nears the bottom of the cylinder it is cushioned by the air which has been drawn in from the surrounding chamber, and that air is forced back into the chamber through the poppet valve V. The degree of cushioning can be accurately adjusted by means of the screw S.

196. REVERSING GEAR. Wolf type. E, is the eccentric;

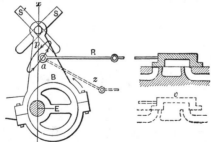

B, eccentric strap and arm; p, a pin sliding in the link, S, which is moved to the position S' for reversing; a R, valve rod connected to the eccentric arm at a. The elliptic line, p, shows the range of the valve motion and swings to the vertical with the link and moves the valve within the range of its lap.

197. Valve just opening, forward.

198. Valve just closing, reverse.

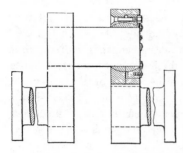

199. FLEXIBLE CRANK for marine shafting. The crank pin is fixed in one side and swiveled in the other side of a double crank, as shown, giving flexibility to a line of shafting in marine engines.

200. FLEXIBLE COUPLINGS for marine shafting. A ball bearing between the sectional ends of a line of shafting. In order to reduce

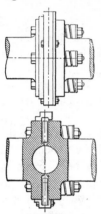

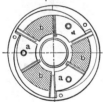

friction to a minimum, a parallel piece made of suitable material is placed between the driving ahead faces of the jaws, *a*, on the driving shaft, and the driven ahead faces of the jaws, *b*, on the driven shaft. These pieces are lipped under the jaws at the bottom or inner end, to prevent them flying out while in motion. For the purpose of taking up the backlash and compensating for any wear that might occur on the driving ahead faces of the jaws, adjustable pieces made in wedge form are fitted between the driving astern faces of the jaws, *a*, on the driving shaft, and the driven astern faces of the jaws, *b*, on the driven shaft.

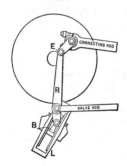

201. Longitudinal section, showing ball bearing, overlap of the jaws, wedges, and volute cap.

202. Shows the alternate jaws, wedges, and the volute cap for tightening the wedges.

203. NOVEL VALVE GEAR. The crankpin arm is pivoted to the lever R at E, and to the link block B, and also to the valve rod as shown. The motion of the valve is controlled and reversed by rocking the link L.

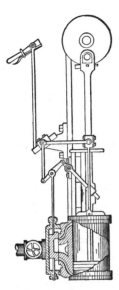

204. REVERSING GEAR without eccentrics. The valve stem is connected to the middle of a short link, one end of which is pivoted to the crosshead bar, and the opposite end to the radial bar, which in turn is pivoted to the link block. The latter member consists of a block of iron grooved to fit the inclined link or reversing bar and having suitable shoes for taking up wear. This block receives motion from a somewhat similar block, which slides on the connecting rod; the block being held in the proper horizontal position by means of a radial rod pivoted to it and to the cylinder. The crosshead bar passes through a sleeve block carried by the crosshead, which is fitted with shoes to take up wear. It will be seen that the crosshead bar imparts a horizontal movement to the valve stem, which movement is equal to the lap and lead of the valve.

205. FLOATING VALVE GEAR or reversing ram for marine engines. The floating lever g is here connected to the crosshead at k.

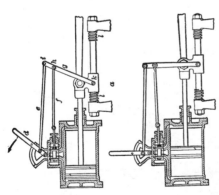

The rod f is hinged at h to the floating lever, and connects it with the valve stem. The rod e is hinged at i to the floating lever, and connects it with the reverse lever d. Then, the piston being stationary, the floating lever swings around k as a fulcrum, and the valve is forced to the left. This valve is an indirect valve, that is, it takes steam at the center and exhausts past its outside edges—just the reverse of the ordinary D slide valve. The lower end of the floating lever moving with the crosshead, it tends to swing around i and thus return the valve to its mid-position. Should the piston creep in either direction, the valve gear will automatically return it to its proper position. To prevent shocks due to a sudden movement of the reverse lever, buffer springs *l, l* are provided, which gradually bring the moving parts to rest.

In both gears shown, suitable stops in the valve chest prevent the valve from being moved beyond the positions required for a full opening of the ports.

206. Shows valve on center for stop motion.

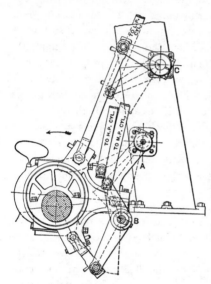

207. TRIPLE EXPAN-SION VALVE GEAR with single eccentric. A, eccentric strap stay arm, which also operates the high-pressure valve rod. B, bell-crank rock shaft that operates the medium-pressure valve rod, linked to eccentric arm. C, rocker arm, shaft, and bell-crank connection by link to the eccentric and to low-pressure valve rod. (At Edison Electric Station, New York City.)

208. WALSCHAERT'S VALVE GEAR as applied to a compound locomotive. The crank-pin arm operates the motion of the

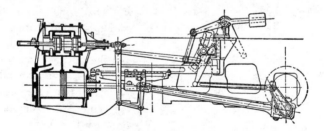

slotted link. The valve-rod block and rod is balanced by a weight on the rock-shaft arm, and operated by a lever connected to the third arm. Valve lead is made by the crosshead arm link and lever connected to the valve rod and link-block rod. Italian railway.

209. REVERSING GEAR. The eccentric is provided with a curved rack near its periphery which meshes with a small pinion. The

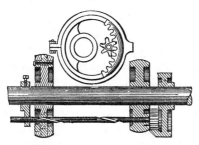

pinion is secured to the end of a shaft provided with a groove disposed spirally for a portion of its length. The shaft is journaled in two collars or flanges keyed to the main shaft so that the small shaft lies parallel to the engine shaft. A third collar slidably mounted on the engine shaft is prevented from turning by a suitable key, this collar carrying the strap to which the reversing lever is connected. A pin in the latter collar engages the groove in the smaller shaft and when this collar is shifted sidewise the pin causes the smaller shaft to revolve, which turns the eccentric around on the engine shaft and thus shifts the position of the valve.

210. Longitudinal section, showing the spiral grooved shaft and pinion.

211. ENGINE STOPPING MECHANISM. If the governor belt breaks, the weight N will drop, and through the system of levers

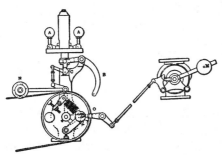

and links throw the bell-crank lever B so as to shift the safety blocks on the knock-off cams of the valve gear and prevent the valves from being opened by the grab hooks.

An auxiliary device is also provided to act in the case of racing. This consists of a small centrifugal governor of the shaft type mounted in the belt wheel of the main governor. The weight w of this auxiliary governor is provided with a lip, which, in the event of abnormal speed, will be thrown outward so as to engage with a small lip, O, on the end of the rocker arm shown. The other end of this rocker arm is connected to a latch which normally holds the throttle open. When the lip on the governor weight w engages the projection O, this latch is thrown, allowing the weight M to close the throttle.

212. SHIFTING ECCENTRIC for stopping or reversing engines. A slotted sleeve sliding on the shaft with wedge-shaped wings that pass through corresponding slots in the eccentric, move the eccentric to the center and reverse by the longitudinal movement of the sleeve and wings. The yoke lever and slotted collar control the movement of the sleeve and wings between the stop collars.

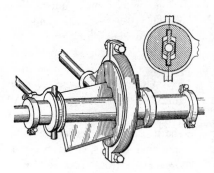

213. Section of eccentric, sleeve and wings.

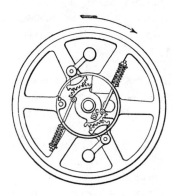

214. SECTOR GEAR GOVERNOR. Two balls on bell-crank sectors with their teeth meshing in a central double sector, to which is attached the compression springs, which are adjusted to the proper set of the eccentric. *b* and *c* are the pivot connections with the eccentric.

215. DASHPOT GOVERNOR. The eccentric is mounted on a plate G, pivoted at P, and is connected to E B, No. 1, and E B, No. 2, by connecting rods in such a manner that the action of centrifugal force in throwing the weights B B outward cause the center of the eccentric to swing toward the center of the shaft. The springs pivoted at K rock against the centrifugal force and hold the weights in a determinate position for each speed. The dashpot simply restrains the motion when too rapid and tends to prevent racing.

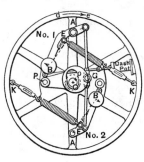

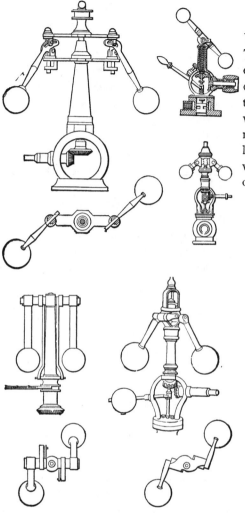

216. CENTRIF-UGAL GOVERNORS. There are patents for several hundred of this type of governors, of which this and the preceding volume of mechanical movements represent the leading models, most of which are practically obsolete.

217. } Slot cam joint
218. } governor.

219. } Crank-pin gov-
220. } ernor.

221. } Adjustable gov-
222. } ernor.

223. Straight-arm governor.

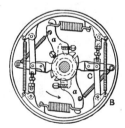

224. FRICTION POWER CONTROLLER. Wick's patent. Transmits only the amount of horse-power it is set for. The power is given to the pulley, B, by the arm *a*, helical springs, and friction sectors. The sectors are thrust in contact with the pulley by the adjustable links C, cams, and thrust bars.

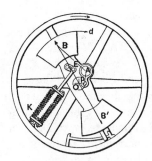

225. INERTIA GOVERNOR. The weights B and B′ are balanced on the center line of the shaft arm, which is pinioned at A to the fly wheel or pulley and to the eccentric at *p*. The spring K holds the weights in normal position, their range of motion by differential momentum from variable speed of the engine being limited by the stop on the rim of the pulley.

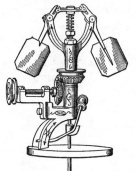

226. FAN GOVERNOR. In which air resistance modifies the centrifugal action of the fans for regulating a gear train and brake. It is an early form for regulating steam engines as shown in the cut. Wing governors are used for regulating gear trains in clocks, music boxes, and revolving window show frames.

227. ADJUSTABLE GOVERNOR. King type. The balls are attached to the shaft by springs and linked to the head and valve spindle, which are drawn down by the centrifugal action of the balls. The regulation is made by the small helical spring and lever. The action is direct through the spindle to the throttle valve.

228. MARINE GOVERNOR. Porter type. A cone pulley with screw-belt shipper for close adjustment of speed. Balls are jointed to the shaft arms with link connections to the sliding collar with resisting spring. The collar carries a central rod to a bell crank and to the throttle valve.

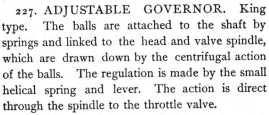

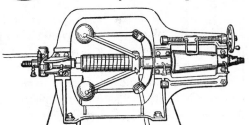

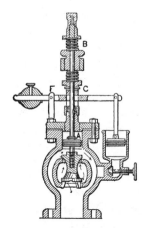

229. DIFFERENTIAL PRESSURE REGULATOR. A supplementary piston and counterweighted lever pivoted at **F**, gives a close adjustment of differential pressures. The steam piston at A is connected with the high pressure side and is balanced by the spring at B, while the supplementary lever is attached to the valve spindle by the block and pin at C.

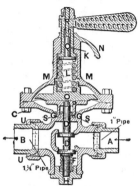

230. BALANCED PRESSURE REGULATOR. Gold type. D, balanced valve. O, Low pressure regulating disk and diaphragm. L, counter-balance spring. Q, adjusting plunger. F, contact spring to keep the plate P in contact with the rubber diaphragm. N, locknut handle. Other parts are self-explanatory.

231. SELF-CLOSING STOP VALVE. The piston on the valve stem has a larger area than the valve disk. The valve is held

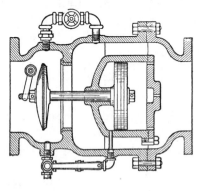

open by the relief from pressure through the by-pass and three-way cock. The dropping of its lever by a lanyard, closes the relief and gives the rear side of the piston the full steam pressure, quickly closing the valve. The by-pass valve at the top is for equalizing the pressure and allowing the valve to open by means of the relief at the three-way cock.

232. REVERSING GEAR for a steam engine. The figures show a side elevation of the reversible eccentric, with handwheel for oper-

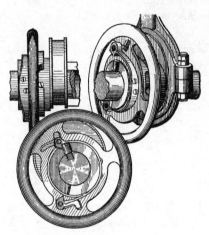

ating it, a front elevation showing the engine shaft in cross section, and a perspective view illustrating the application of the invention to an upright engine. The eccentric is formed with a hub having shoulders to engage a stop pin on the shaft, in combination with an operating wheel placed on the hub of the eccentric, and having a limited rotary motion thereon. The eccentric has a limited independent motion upon the shaft, and the handwheel has a rotary motion independent of the eccentric, combined with spring catches arranged to lock the handwheel to the shaft.

233. Perspective view.

234. Front view.

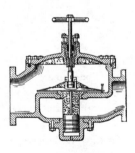

235. NOVEL REDUCING VALVE. Holly type, having a large area and lap of a flat valve disk. The relative difference of pressure is regulated by the free hanging weights under the disk, while excessive back pressure tends to close the valve by pressure on the large area of the back of the disk.

The wheel and screw spindle is to close the valve when required.

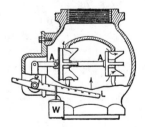

236. DIFFERENTIAL EXHAUST VALVE. For regulating the back pressure on the engine in exhaust steam-heating systems.

The two-winged valves are nearly balanced, requiring only a small weight to balance them and prevent chatter of the valves.

237. AUTOMATIC QUICK-CLOSING VALVE. The bonnet piston C has a larger area than the valve disk and communicates with the steam pressure in the main pipe through its hollow spindle. The leakage of steam around the loose-fitting sleeve of the piston at G equalizes the pressure on both sides when the relief pipe is closed.

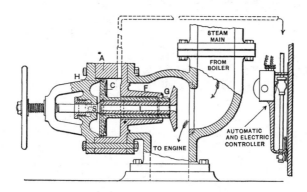

238. The automatic electric controller, shown at the right, has a magnetic dog that disengages a weight which falls against a lever and opens the relief valve and quickly closes the valve by the greater pressure on the rear of the piston.

The screw spindle S closes the valve as an ordinary stop valve. The electric push buttons are placed where needed for emergencies.

239. REVERSIBLE THROTTLE VALVE. In this design an angle or straight way valve may be made convertible by rotating the flange connection of the two parts of the body.

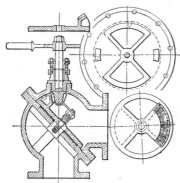

A most convenient design for facilitating repairs. The valve spindle carries a bevel pinion meshing in a sector gear on the valve disk, which opens or closes by a 90° revolution on its face.

240. Plan of valve disk.

241. Sector gear on disk.

242. COMPENSATING EXPANSION JOINT.   Designed to prevent the forcing apart of the ordinary expansion joints in steam pipes.

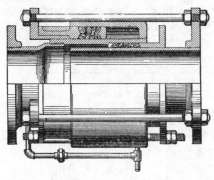

The joint is surrounded by an annular chamber of cross section equal to the steam pipe, in which a tightly packed ring acts as a piston. Steam is admitted to this chamber by means of a by-pass. The tendency would be to force out the piston, and so draw the ends of the pipe closer together, but as the steam in the pipe and in the chamber is of the same total pressure, each force neutralizes the other, and the joint is rendered secure under all ordinary circumstances.   The joints are made of steel pipe and forgings, excepting the glands, which are cast, and the first cost is very little greater than that of an ordinary joint.

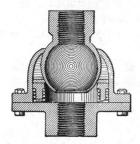

243. FLEXIBLE BALL JOINT.   The space between the ball and shell is filled with an elastic lubricated packing held in place by an annular follower and springs.

244. BALANCED EXPANSION JOINT for steam pipes Smith pat.   Referring to the cut, it will be noticed that the inner tube

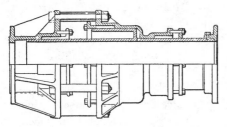

has an increased diameter or ring about halfway along its length.   This forms a shoulder or piston at the end next to the bottom of the large stuffing box casting.   The other end of this annular piston or ring is open and is steadied by the gland.   In the inner tube below this ring there are holes which admit steam from the main, back of the shoulder. As the exposed area of the shoulder or piston is equal to the area of the steam main, the pressure in the main is equalized. As the stuffing box is tied to the other end of the joint by long bolts the entire line of pipe is in a state of equilibrium so far as the end pressure is concerned. The expansion due to heat is provided for by a liberal space for end play at the cast end of the joint.

245. UNIVERSAL FLEXIBLE PIPE JOINT. The internal construction shown by the section shows how contact of the gas or

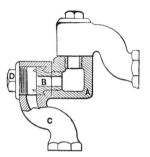

fluid which might corrode the wearing surfaces is prevented and at the same time insures that the movement of the parts shall be smooth and free.

The material is cast iron except the piece B, which is bronze. The body A is threaded for B with a slightly tapered thread, so that when B is screwed home there shall be no leakage between the parts. The under side of the head of B is formed into a conical seat which makes a steam or gas-tight joint with C and the flat faces between C and A are round together also, making an additional safeguard against leakage.

246. CARGO ELEVATOR for loading and unloading ships. Otis type. Steam driven by a double engine geared to a shaft on

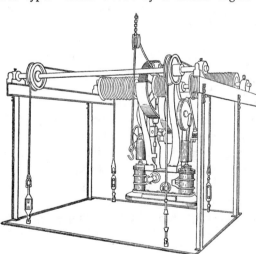

which two double drums are fixed.

Four cables from the drums are attached to the corners of the platform with turn-buckle adjustment.

Automatic adjustment for stopping at any deck for loading or unloading. Capacity two tons at a speed of 100 feet per minute.

247. FACTORY HEATING FROM WASTE GASES. Cold

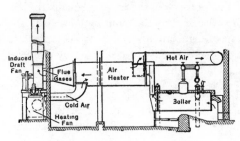

air is blown through the annular chamber, between the boiler and the chimney, by a fan, heated and distributed for heating rooms.

Additional draught may be given to a chimney by a high-pressure blower and jet nozzle in the uptake.

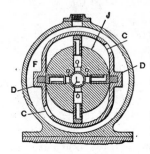

248. ROTARY ENGINE. Takes steam through the shaft L. The abutment pistons, *o, o, o, o,* are pushed outward by the steam pressure and have ports that are opened after passing the closure blocks D, D. The steam ports are closed by pushing in of the abutment pistons at the exhaust ports C, C. E, exhaust jacket, F, exhaust space.

249. REVERSIBLE ROTARY ENGINE. On the driving shaft, within the cylinder, is secured the hub of a wheel containing a

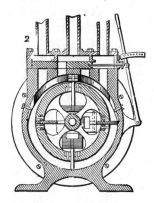

series of four pistons fitted to slide in the rim of the wheel, the opposite pistons being connected in pairs at their inner ends by a slotted frame through which the shaft passes, so that the pistons have free radial movement, one moving inward as the other moves outward, and *vice versa*. The outer ends of the pistons engage the inner surface of part of the cylinder and part of an abutment in the cylinder. The abutment is made in two parts, bolted at their outer ends to the cylinder, and connected with each other at their inner ends by bolts and intervening packing strip, and the abutment serves to press an outermost piston inward, so that its opposite mate slides outward into contact with the peripheral inner surface of the cylinder.

250. ROTARY ENGINE. Harrington type. The disks have a bearing surface of several inches on each other, preventing the passage of

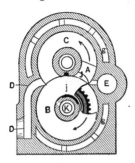

steam between them. An end elevation partly in section, showing the piston A, and the abutment disk B, in position at the instant of taking steam through a port from the valve chamber E. The piston disks and gear are attached to the driving shaft, and the abutment disks and gear are attached to the shaft K. These shafts run in taper phosphor-bronze bearings, which are adjustable for wear or other causes by screw-caps. The whole mechanism is kept rigidly in place by the flanged hub. The flanged heads project through the cylinder head, touching the piston disk, and thereby prevent any end motion of the shaft.

251. ROTARY STEAM ENGINE. French design. The engine consists especially of a jacketed cylinder, C, in the interior of which

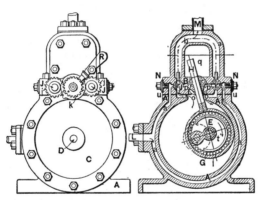

rolls a piston ring, G, carrying at its upper part a partition, H, always connected with a special oscillating piece, r, r, and contributing toward continually dividing the interior space into two compartments, the capacity of one of which varies in inverse proportion to that of the other. Two cocks, J, J, placed on each side of H, serve to establish the admission escapement according to the direction of running it; and the maneuvering is very easily effected by means of a simple handle that actuates a toothed wheel that gears with the two cocks. The motion of the piston ring is transmitted to the driving shaft, D, through the intermedium of two symmetrical cams, E, united at their center by a rod and nut, which permits of regulating their distance apart. The joint between these two cams, placed in the axis of the motor, therefore constitutes a more or less open channel in which

is placed a series of tempered steel balls that roll upon a corresponding path arranged in the interior of the piston ring. Two cheeks
traversed by the driving shaft close the cylinder at the sides, and a
perfect tightness between these cheeks and the lateral faces of the
piston ring is obtained.

252. Vertical section, showing details of the parts.

253. ROTARY STEAM ENGINE. A cylindrical piston A,
with wing abutments C, C. A double cam block H, made adjustable

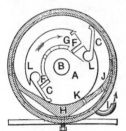

by a set screw and with exhaust ports at K
and I. A steam port through the cylinder
cover and a curved passage in the cover shown
by the dotted lines, so that the abutment pistons take the full steam pressure through the
sector passage, from F to G, and expanding
through about one-quarter of a revolution of
the piston.

254. ROTARY ENGINE. Recesses are formed in tne piston
having an S-shaped partition between them, the recesses opening at

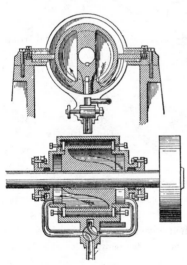

opposite ends of the piston into
the steam chests, and by means of
ports into an annular space between the casing and piston. The
piston has an eccentric portion
which has a perfect contact bearing with the inner wall of the
casing by means of a yielding
block in a recess on its periphery,
the eccentric also acting alternately
to press back abutment blocks
adapted to slide on antifriction
ball bearings in recesses in the
arms of the casing. The steam
supply pipe connects with a passage communicating with a circular
chamber in which is a rotary valve,
by means of which steam may be directed into either of the branch
pipes connecting with the steam chests at the ends of the piston, the
arrows showing the direction of the steam when admitted into the

right hand pipe. The valve chamber also connects with a steam discharge pipe, the valve being turned by means of a handle or wheel to direct the steam into one or the other of the branch tubes, when the opposite tube will form the outlet pipe for reversing the engine.

255. Longitudinal section, detailing the parts as above described.

256. PENDULUM COMPOUND ENGINE. This is a compound engine of the pendulum type, the upper or high-pressure

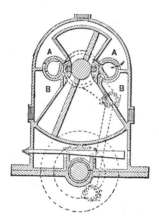

cylinder being surrounded by the steam chest A A. When the pendulum is in the position shown the live steam is admitted to the right of the high-pressure blade, as shown by the arrow. Meanwhile the steam on the left of the upper blade is exhausted from that chamber to the left of the low-pressure blade. The spent steam on the right of the low-pressure blade is exhausted into the right hand exhaust chamber B. The manner of transforming the motion of the pendulum into rotary motion is plainly shown in the figure. It will be noticed that the upper end of the connecting rod has a reciprocating circular motion.

257. ROTARY ENGINE. The piston is of the usual drum pattern mounted eccentrically upon the main shaft. The abutment is

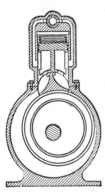

carried by a cylindrical guide block and rests upon the piston, being free to oscillate upon the pin. Steam is prevented from escaping from the steam to the exhaust port by the abutment, which has a sliding contact with the partition in the cylindrical guide. The upper end of the guide terminates in a piston which works in the cylindrical upper portion of the engine casing, and is normally pushed downward by a spring above it, the downward pressure of the spring, together with the action of the steam on the oscillating abutment, being adapted to keep the latter firmly in contact with the piston drum. The engine is reversed by changing the inflow of steam from one side of the partition referred to, to the other, which operation also reverses the exhaust openings.

258. ROTARY PISTON ENGINE.   Has a casing or frame in which is an annular groove or cylinder.   In this groove is fitted a pis-

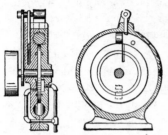

ton which is carried by the piston disk as shown.   A sliding abutment is raised and lowered by means of a cam on the engine shaft and a cam rod. The revolving shaft and cam causes the abutment to rise and fall at the proper point in the travel of the piston.   In the section the abutment has just begun its downward stroke, forming as it does the cylinder head.   The piston disk is provided with a radial groove which communicates with an annular groove shown.   Steam is admitted to the annular groove by means of a slide valve shown by dotted lines and immediately below the shaft.   As soon as the abutment reaches the piston disk, the valve opens and admits steam into the annular groove and thence into the radial groove, the latter conducting it into the space between the piston and the abutment, the steam pushing the piston around in the annular cylinder. The steam is exhausted through the large exhaust opening at the right of the abutment when the piston reaches this point in its stroke.

259. Longitudinal section, showing steam connections.

260. OSCILLATING ROTARY ENGINE.   This engine consists of a pair of curved cylinders, P P, a circular piston rod, L, to

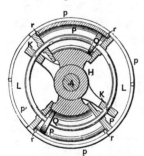

which are attached the two pistons, Q Q, the pistons traveling within the cylinders with a reciprocating rotary motion, and a pair of rocking radial arms, K K, which transmits the motion of the piston rod to elliptical gears (not shown in the engraving), which "controls the motion and transmits the power of the engine."

The cylinders with the steam pipe, p, are carried by a bracket, H, which is keyed firmly to the main shaft, A, and rotates with it.   The radial arms, K K, to which the piston rod, L, is keyed, are journaled loosely upon A, and carry an elliptical gear, which meshes with another elliptical gear carried upon a countershaft.   When steam is admitted to the cylinders through the ports, *r*, at either end of the cylinders, the difference in the diameters of the elliptical gears at the point of contact causes the main shaft elliptical gear, with the attached parts, to rotate.

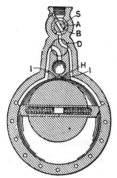

261. ROTARY ENGINE. Casaday's patent. A reversing rotary with adjustable cut off. A is the cut-off plug within an adjustable rotable cylinder and operated by an arm and connecting rod to the eccentric or the plug rotated by sprocket and chain. D, reversing plug. H, abutment block, cushioned by steam through the passages I, I.

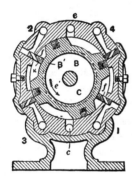

262. REVERSIBLE ROTARY ENGINE. The piston when revolving in the direction of the arrow takes steam by the throw of a two-way cock; steam entering through the diagonal slots in the abutment pieces at *f, f*, and exhausting through the ports *e, c*. Reversed by throwing over the steam-cock opening to the passages to the opposite ports in the cylinder. 1 and 2 are forward and 3 and 4 are reverse ports.

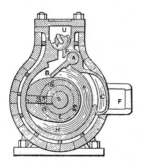

263. ROTARY ENGINE. Hodson type. The valve U is operated by a cam on the shaft S, to cut off for expansion. A, B, is a riding valve in contact with the elliptic cylinder, which has a packing slide, S, P, concentric with the axis and following the wall of the shell as a packing.

Steam follows at half or less part of a revolution and then expands to the exhaust port, C.

264. STEAM RAM for elevating water. Erwin type. Penberthy Injector Co. Water is elevated by the alternate action of steam and atmospheric pressure. The steam having first driven the water from the ram is instantaneously condensed and a vacuum is formed. A volume of water is then driven into the ram by atmospheric pressure.

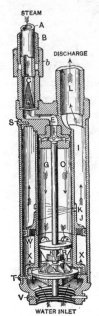

The ram is placed beneath the surface of the water in the well or other source of supply and, before starting, water flows into it by gravity. When steam is turned on it passes through the steam pipe A, nipple C, conical screen D, the main steam port E, and radial steam ports F into the cylinder G. The water is then forced downward through the openings H into the surrounding discharge chamber I, where it passes through the annular check valve J and out of the discharge pipe L.

When the steam reaches the lower end of the cylinder G, it is exhausted through the large openings H much faster than it is admitted through the steam ports F, is condensed in the surrounding discharge chamber I, and a partial vacuum is formed within the cylinder G. The vacuum is made more complete by a spray of water which then rushes inward from the discharge chamber I through the small opening K.

The instant a vacuum is created and condensation occurs the pressure of the atmosphere on the water outside of the ram forces water upward through the bottom strainer. The main check valve N then rises and the valve rod O, which is rigidly attached to it, shuts off the steam at the upper end of the cylinder. A volume of water under atmospheric pressure is at the same time forced upward through the discharge chamber and out into the discharge pipe. A portion of this water, however, passes through the openings, forces up the float R, which moves freely on the valve rod O, and refills the cylinder.

The water under atmospheric pressure having then lost in momentum, the steam acting downward on the valve rod closes the main check valve, and through pressure exerted on the float, again forces water out of the cylinder and through the discharge chamber and discharge pipe.

A covering pipe B surrounds the steam pipe for the distance it is submerged beneath water, to prevent condensation, and is received into the coupling *b*.

SECTION VI.

EXPLOSIVE MOTOR POWER AND APPLIANCES.

Section VI.

EXPLOSIVE MOTOR POWER AND APPLIANCES.

265. THE LIGHTEST GASOLINE MOTOR. Duryea Power Co. type, Reading, Pa. The motor is a 6-cylinder, using gasoline as

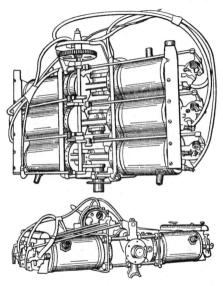

fuel, being of the opposed cylinder type, and working on a 3-throw crank shaft in perfect mechanical balance. As it appears in the cuts it weighs slightly over 200 pounds, or less than five pounds per horse-power. With spark coil, battery, fuel and water tanks partly filled, it weighs 232 pounds, or 5.7 pounds per horse-power. The cylinders are $4\frac{1}{2}$-inch bore by $5\frac{1}{2}$-inch stroke, with bearings of the same size as used in the company's regular automobile motors. Jump-spark ignition is used, having a single coil and commutating the secondary current. The inlet and exhaust valves may be removed from any cylinder head by loosening a single nut. The crank shaft and crank pins are hollow for lubrication purposes.

This motor is believed to be the lightest for its power ever constructed and is another evidence of the mechanical development brought about by the requirements of the automobile.

266. Side view of the motor, showing sparking rod connections with the secondary shaft.

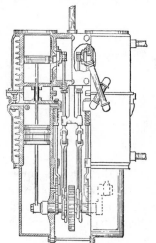

267. COMBINED GASOLINE AND STEAM MOTOR. In this design the piston of the explosive motor is made the crosshead for the connecting rod. A duplex steam engine with a duplex explosive motor as an auxiliary power in which the exhaust of the steam engine may also be turned into the explosive motor cylinder as an additional power and lubricant when the explosive motor is not in use.

268. TWO-CYCLE MARINE MOTOR. Lozier type. The principal features are the throttle valve to regulate the charge from

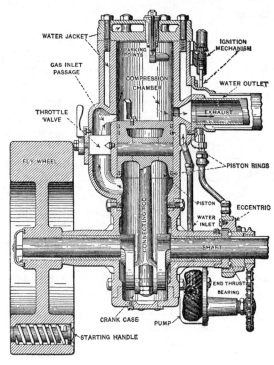

the crank chamber and the operation of the hammer spark break from a cam on the shaft. A rotary circulating pump is driven by chain from the main shaft and the discharge of the water from the cylinder is around the exhaust pipe. The thrust is taken by ball bearings in the cam hub. A throttle valve in the passage from the crank chamber to the cylinder regulates the charge.

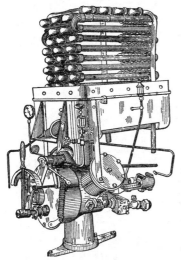

269. ALCO-VAPOR BOILER and three-cylinder engine. Alcohol of low grade is injected into the pipe boiler and converted into vapor under pressure by the heat of part of the vapor burned under the boiler. The three cylinders are single acting on a single crank. Casing of boiler is removed to show its construction.

The exhaust vapor is condensed in a keel condenser and returned to the tank. The boiler pressure gives force to the vapor jets in the Bunsen burners of the furnace.

270. KEROSENE OIL ENGINE. Two-cycle Weiss type. E, D, conical vaporizer inclosed in a shell for confining the lamp flame for starting the engine; *h*, inlet valve with spring to hold it closed subject to the action of the pump *g* ; *e*, pick blade that drives the pump

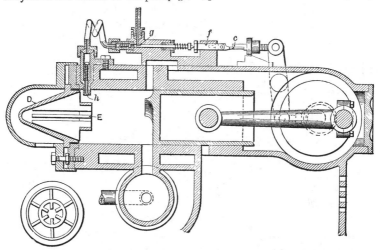

piston *g*, for a measured charge of oil. The hit or miss is regulated by the lifting of the pick blade on the incline of the wedge beneath the collar of the pick blade, which is made adjustable by the nut and screw on the pick blade. The wings of the conical vaporizer are shown in the section.

271. GAS OR GASOLINE ENGINE.   Air-cooled four-cycle
type.   Ribs around the cylinder and on the head.   The novel features
are the long crank-shaft bearing with the supplementary crank, 45, and

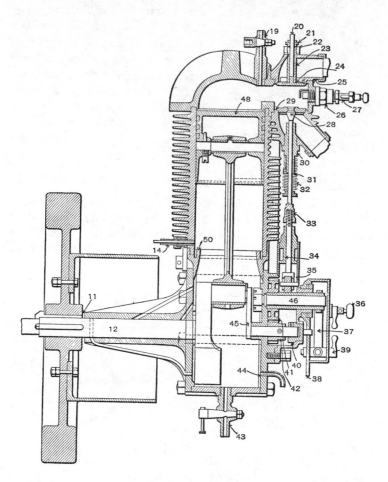

reducing-gear shaft, 46, carrying the cam-roller movement for the
exhaust valve, the spark-breaker cam, and contact bar, 37, and the
regulating screw, 39.   The atomizer or vaporizer connects with the air
inlet at 24, the air cock for starting at 19.   The other parts are self-
explained.

272. BALANCED ENGINE. Explosive motor. Secor type. The charge is fired in the chamber X between the two pistons H H', whose

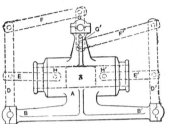

motion is transmitted to the cranks G G', having equal throw and set at 180° apart on the crank shaft.

The pistons are connected by the short connecting rods H H' to the vertical levers D D', which transmit motion to the cranks through the connecting rods F F'.

273. GASOLINE ATOMIZER AND VAPORIZER. Hay type.

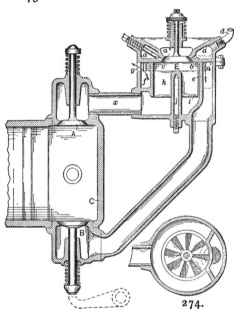

274.

The exhaust is used for heating the walls of the vaporizing chamber by traversing the annular chamber e. A fan, h, is revolved on the spindle, j, by the inrushing air and gasoline through the valve E, which also covers the gasoline inlets on the face of the valve seat and is connected with the annular chamber, a, and pipe, d. The gasoline feed is regulated by the needle valve a. Other details and exhaust passage are shown in the horizontal section, 274.

275. SOOT-PROOF SPARKING PLUG. For gas engine.

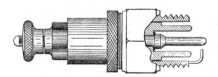

Merger type. An annular projection on the end of the porcelain insulator extends the insulating surface and prevents short circuiting of the electric spark.

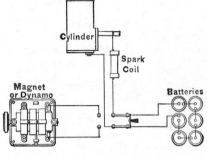

276. IGNITION CONNECTIONS for gas engines. Showing battery cut-off switch of double throw type, location of spark coil, and current breaker on engine. If a jump-spark igniter is used, an induction coil should be substituted for the spark coil.

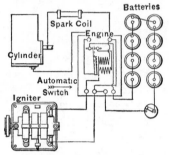

277. IGNITION CONNECTIONS for gas engines, showing a one-point switch to cut out battery and an automatic switch so arranged that failure of the dynamo igniter current turns on the battery by release of the armature of the automatic switch. On restoring the dynamo current, the automatic switch cuts out the battery.

278. MULTIPLE CYLINDER IGNITION. Bosch type. The armature, A, which is stationary, is provided with two windings,

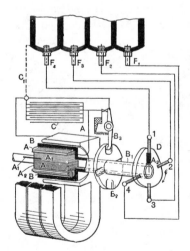

A^1 and A^2, of which A^1 is of stout wire, and corresponds to the primary winding of an induction coil, A^2, being of fine wire and corresponding to the secondary. The changes of magnetism in the armature core, which give rise to the current, are produced by the rotation of a soft iron sleeve, B, which partially surrounds it, and is integral with the hollow shaft, B^1, which also carries the notched disk, B^2, and the high-tension distributing disk D. One end of the winding, A^1, is grounded on the shaft of the apparatus, and the secondary winding forms a continuation of the primary. The other end of the primary winding, A^1, is led to one side of the contact-breaker, B^3, and to one terminal of the condenser, the other terminal of the condenser and the moving arm of the contact-breaker, B^3, being grounded. The sleeve, B, is slotted, and when the slots come opposite the poles of the field magnet, the armature receives magnetism from the field magnet, and is deprived of it again as the slots pass around, and a powerful current is consequently induced in its windings. The contacts of the contact-breaker, B^3, are normally held together by the action of the disk, B^2, and during these periods the low-tension winding, A^1, is closed on itself, so that a powerful current flows through it at the moments when the magnetism of its core is being varied by the rotating sleeve B. When one of the notches in B^2, which are steep on one side and beveled on the other, come under the lower end of the contact lever arm, B^3, the latter snaps back, owing to the action of its spring, separates the two contacts, and breaks the circuit of A^1. This produces a high-tension current in the secondary or fine wire winding, A^2, the condenser, C, increasing the effect. As the secondary winding is connected to the primary as described, and as it is grounded through it, successively connecting the central rods of the sparking plugs, F^1, F^2, F^3, F^4, to the opposite end of the secondary, A^2, causes sparks to pass in the four cylinders at the right moments, the tension or voltage of the primary and secondary being added to one another. The distribution is effected by the commutator, or distributor, D. This consists of the rotating disk, D, carrying the metal plate, A^2, which is in conducting connection with the insulated end of the secondary winding A^2. As the disk revolves, this metal plate makes contact successively with the fixed brushes 1, 2, 3, 4.

279. GASOLINE MOTOR STARTER. A starting wheel B, with oblique saw teeth, is fixed on the motor shaft A. A sprocket chain C, C' is wound on a drum containing a coiled spring D, so arranged as to rewind the chain with a stop J, so as to allow it to hang free from the ratchet wheel when the finger loop at E is dropped to the eye in the vehicle floor. G, sheave, K, slotted guide plate, F, lanyard. To start, pull on E to catch the chain in the teeth of the wheel and with a jerk set the wheel revolving, and. if necessary, repeat.

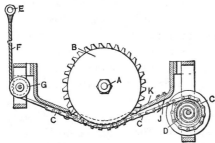

280. MUFFLER FOR EXPLOSIVE MO-
TORS. Thompson type. A cylindrical chamber
with a hooded spreading inlet pipe ; a deflector on
the exit pipe, by which the exhaust puffs are expanded in the cylinder and issue in a nearly constant stream.

Other types of mufflers have strong wire gauze cylinders within the drum so arranged as to break the
impact and disperse the exhaust before it leaves the
outer shell.

281. EXHAUST MUFFLER for gas, gasoline, or other engines.
A perforated exhaust nozzle within an open end pipe of larger size.

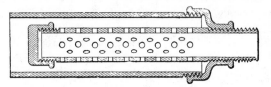

Its construction is
shown in the cut.

The outside or
shell of all mufflers
should be felted with
asbestos.

281A. CRANKER FOR GASOLINE ENGINES. Among the
latest self-starters for gasoline engines is a mechanical cranker which

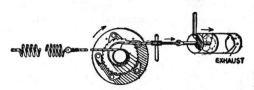

imitates the accelerated
speed of the hand
crank. It is simple in
design and can be op-
erated from an auto-
mobile seat. The crank-
er itself consists of a disk mounted so as to revolve with the engine
shaft and carrying a number of pawls which engage the teeth of a
small wheel which turns loosely on the same shaft. A cable, attached
to the piston rod of a cylinder for compressed air or gas, passes around
a spiral drum on the toothed wheel, and is fastened to a spiral spring
at the other end. When compressed air is turned into the cylinder,
the piston unwinds the cable from the drum, the toothed wheel
catches a pawl on the disk, and the engine shaft is revolved; the spring
then returns the cable to its former position. The spiral shape of
the drum gives the accelerated speed of hand cranking. When the
engine is running, centrifugal force swings the pawls away from the
toothed wheel.

SECTION VII.

HYDRAULIC POWER AND APPLIANCES.

Section VII.

HYDRAULIC POWER AND APPLIANCES.

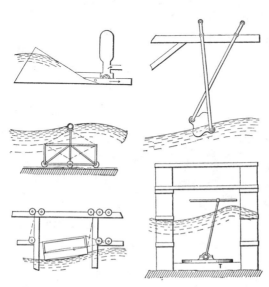

282. WAVE MO-
TORS. Waves oper-
ating a hydraulic ram.

283. Waves oper-
ating swinging levers.

284. Waves push-
ing a vertical surface.

285. Waves lifting
a float.

286. Waves swing-
ing a hinged blade
anchored on the bot-
tom.

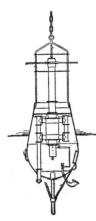

287. FOG-HORN BUOY. A float anchored
at the edge of banks with an air pump operated by
the waves. The action of the sea is utilized in such
a manner as to blow desired blasts through a fog-
horn by means of the compression and release of air
into and from a suitable air-tight chamber forming a
portion of the buoy, this chamber being charged by
means of a pump actuated by the movement of the
sea.

288. ORIENTAL IRRIGATION WORKS of "ye olden time"

and yet in use. An ingenious device for the age in which it originated.

The ox-hide bucket and spout drawn by oxen with lines D and E so arranged that the spout line was stopped at H and the bucket raised to automatically empty the water into the conveying trough.

289. CENTRIFUGAL PUMP.

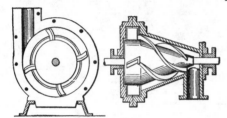

Spiral wings on a conical drum act as a gradual feeder to the main wings at the large end of the cone.

The two sets of wings or blades are inclined at opposite angles to counteract end thrust. Wenzel patent.

290. Longitudinal section, showing both sets of wings.

291. VALVELESS ROTARY PUMP. The piston has a helicoidal form. It is fixed upon an axle, which, running in stuffing

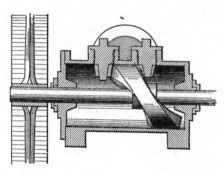

boxes, passes through a cylindrical pump chamber closed at both ends. Two rollers enter this chamber at right angles with the axle and bear against the oblique faces of the piston. It results from this that if the latter is revolved, a backward and forward motion will at the same time be given which will have the effect of producing on each side a suction and compression. Two tubes are placed upon the pump chamber at a slight distance from the rollers. If we examine the operation of the piston, we see that when it is at one end of its travel the tubes are partially covered by it, and if they are sufficiently wide, they communicate with each other on the same side of the piston. But as the latter moves away from the extreme points, the tubes are separated and become independent. There is then a compression on the side toward which the piston is moving and a suction on the other.

292. ROTARY PUMP. A represents a cylindrical casing, provided on opposite sides with chambers, B B, containing sliding abut-

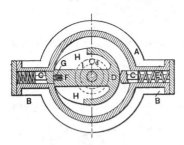

ments, C, which are pressed forward toward the piston, D, by means of the coiled spring E. The piston is constructed in the form of an oval disk mounted eccentrically upon a shaft. The longer end or side of the piston revolves in contact with the interior of the cylinder, and is chambered on opposite sides of its periphery, leaving a partition, F, which is provided with a spring packing, G, at its edge. The chambers, H H', are entirely separate from each other .when the piston is inserted in the casing. One chamber communicates by means of an opening, *d*, through the side of the valve, with an annular groove, I, in one head of the cylinder, and the other with a similar groove in the opposite head, which lead respectively to the induction and eduction ports.

293. CENTRIFUGAL PUMP. German type. The revolving

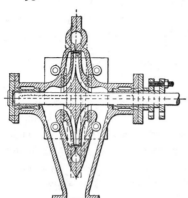

disk receives the water on each side near the shaft in curved channels, and discharges through openings in the periphery of the disk opposite to a continued slot in the casing. A corrugated closure of the shell and disk near the shaft prevents back flow of the water escaping over the periphery of the disk, thereby adding to the efficiency of this class of pumps.

294. RIVER MOTOR. Wheels and chain paddles set in a

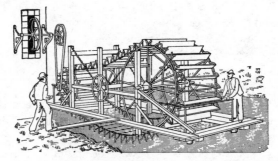

frame on piles, or on floats in a stream, shows increased power by the large number of submerged buckets.

295. FLOATING MOTOR FOR RIVERS. A wheel of the windmill type is hung within a bell-mouthed case, which may be

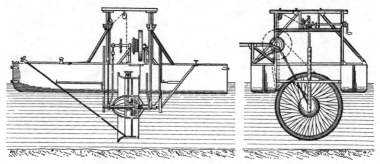

lowered or drawn up to clear the varying depth of a stream and to utilize the full value of the increased velocity at mid depth. Power is transmitted to the shore, or may be used on the floats for pumping water for irrigation.

296. Cross section, showing wheel and frame.

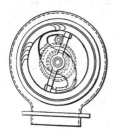

297. WATER MOTOR. A curved bucket-rim wheel revolves around a fixed double jet. The control of the jets is made by wedges on an arm to which is attached a gear. A pinion meshed in the large gear, with shaft extending to outside of case, moves the arm and wedges.

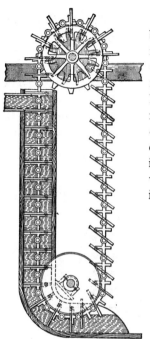

298. WATER MOTOR. Chain-bucket system. Consists of a series of feathering floats hinged to chains running over sprocket wheels and guided in grooves in the sides of a casing. By the use of a large number of inclosed buckets closely fitted, an efficiency of 90 per cent is claimed. This claim is doubted in consideration of the friction of the many buckets in the tube, their loose parts, and the two wheels. A matter too often neglected by inventors. English patent.

299. 1000 HORSE-POWER TURBINE. Swiss type. A single nozzle set on the inside of a curved bucket wheel. The nozzle is broad

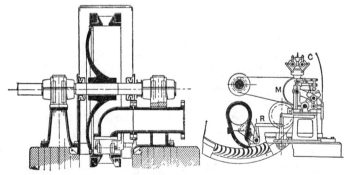

to match the buckets. The water flow is governed by the opening or closing of a sector slide valve controlled by a fly governor.

300. Elevation, showing position of nozzle and buckets.

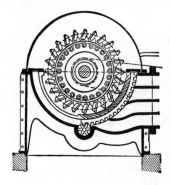

301. MULTINOZZLE TURBINE or impact wheel. German type. The nozzles are in a segment and closed for regulation of power by a sliding segment on the outside, operated by a pinion with a controlling wheel outside the case. The water strikes the face of the buckets and is discharged at their sides.

302. **VALVE MOVEMENT.** Duplex pump. Knowles type.

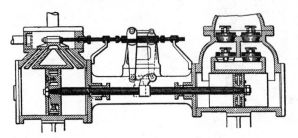

A rocker arm linked to the piston rod of each side of the pump operates its opposite valve.

303. VALVE MOVEMENT. Single steam pump. Knowles type. Freedom from a dead center is secured by the use of the

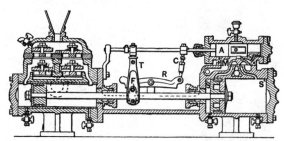

auxiliary piston A, which works in the steam chest and drives the main slide valve M. This main valve is of the B form and moves on a flat seat ; it has on top a stem which fits into a recess in the chest piston A. In addition to this it has on each end a small lip which alternately covers and uncovers a small fifth port S, which enters the cylinder at each end near the head. In operation, the steam piston runs over the main port, and this small fifth port being closed by the above lip, a cushion is obtained. When the main valve M is reversed the lip uncovers the port and admits steam, which starts the piston back easy until the main port is uncovered ; the pump thus changes its stroke very smoothly.

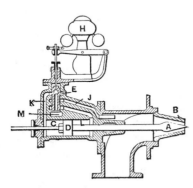

304. IMPACT WATER-WHEEL GOVERNOR. A conical valve, A, in the nozzle, B, controls the volume of the jet and is operated by the piston, D, in the hydraulic cylinder C. E is a piston valve and ports, taking water pressure through passage, J, and delivering pressure to the cylinder at C or L, as controlled by the governor.

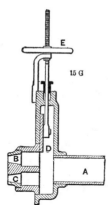

305. DOUBLE-PORTED NOZZLE and valve for impact water wheels. In graduating the flow of water by closing one nozzle the full velocity of the water jet may be retained and the wheel operated at full speed with half the power.

The normal speed of wheels of the impact class is at one-half the velocity of the water at their peripheries for best effect.

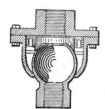

306. FLEXIBLE BALL JOINT, which can be packed with any kind of packing held in place by rings and springs.

This design prevents sand or grit getting into the joint bearings and causing leaks.

Tubbs patent.

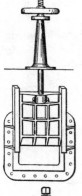

307. IRON SLUICE GATE. Type of designs used for water works, power plants, and irrigation systems. Easily bolted to a wooden flume or anchored to masonry. A water works type for the largest gates.

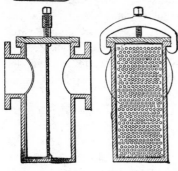

308. BASKET STRAINER. A perforated plate slid into a cylindrical chamber with cover and yoke in a line of suction pipe. Easily removed for cleaning.

309. Section showing perforations in the strainer plate.

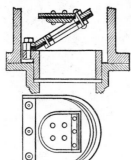

310. DOUBLE-BEAT FLAP VALVE. The reversed flap on the back of the main flap is a great relief to the strain and wear of the valve. Gives a full flow with small lift.

This design is well suited for very large pump valves, section and plan.

311. WATER STILL. A tin-lined copper still and worm of two gallons capacity will yield about six gallons of distilled water per day. Water is fed to the inverted siphon at D, and is sealed against the low steam pressure by the drop of the bend at B. Still should not be over half full for best effect.

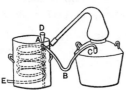

312. WATER-PRESSURE REGULATOR. Used in high-pressure service pipes. The diaphragm *x* and plunger *s* operate the

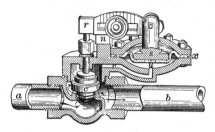

valve *c* through the lever *r*, and the relative pressures are regulated by the movable fulcrum *e*. *a*, high-pressure pipe; *b*, low-pressure pipe; *v*, passage to diaphragm from low pressure.

313. VENTURI TUBE AND MEASURING METER. The differential velocity of the water in the main pipe and in the throat of the double conical tube produces a differential pressure in the small tubes with their mouths turned in opposite directions, which is used for registering the amount of water flow in the main pipe by the flow through the small pipes. The measurement is made by a meter.

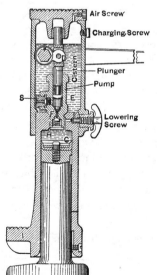

314. HYDRAULIC LIFTING JACK. The shell of the jack extends over and nearly to the foot of the ram to enable a lift at both head and foot of the jack. The small plunger and valves are operated by the lever and arm D, for lifting, and the by-pass valve serves for lowering a load or closing the jack by its own weight, which will send the oil back to the cistern. S, suction valve; F, discharge valve; G, leather cup packing.

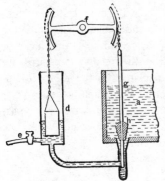

315. UNIFORM FLOW OF WATER from a variable tank head. *a*, a tank with variable head of water; *d*, small tank with float connected to sectored arm *f* and to the conical valve *h* in the tank *a*. When well-adjusted will give a uniform head in *d* for the outlet at *e*.

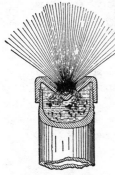

316. NOVEL SPRAYING NOZZLE for cooling water by contact of the spray water with the surrounding air. The apertures in the concave cap give a slightly spiral direction to the jets, which gives them a rotary motion and disintegrates the water into a fine spray.

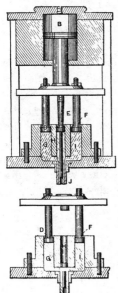

317. HYDRAULIC PRESS for making tin-lined and plain lead pipe. The piston B of a hydraulic press serves to operate the central plunger E and the annular plunger DF. The former forces out the tin contained in the central cylinder H, and the latter the lead in the annular cylinder G I. J is the mandrel. The machine admits of being arranged for making pipes all of one metal, as in the lower figure, 318.

There are several modifications of these lead-pipe machines in use. One on similar lines to the upper figure is for covering electric cables in which the cable is fed through in place of the central plunger.

318. Section for making all lead pipe.

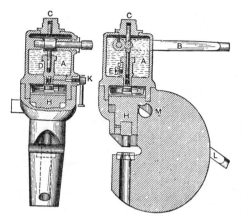

319. HYDRAULIC PUNCH. A small plunger in the cylinder D, with inlet valve E and discharge valve below, is operated by the lever B, to give great pressure to the ram H, to which is attached the punch. On opening the by-pass valve K, the ram is lifted by the lever L and revolving wedge M, pushing the oil back into the reservoir A.

320. Cross section of the hydraulic punch, showing lever action.

321. FIRE EXTINGUISHER. Grinnel sprinkler type. Each sprinkler is calculated to supply an area of 100 feet. The valve, a

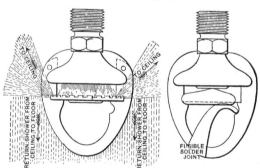

leaden disk affixed to the center of a larger disk of brass, is held up against the valve orifice by a system of two curved levers, the lower of which is secured by fusible solder at its lowest point to a light metal frame. The valve seat is itself made elastic by the device of fixing it in the center of a diaphragm of thin, hard metal, perforated for that purpose, and the pressure of the water upon the diaphragm keeps it tight against the valve. The larger disk attached to the valve disk serves as a deflector. When the solder is melted, the levers fly apart, and the valve and deflector drop about $\frac{1}{2}$ inch, leaving space for the water to escape. It dashes against the disk, which is notched and slightly dished at its edges, and is then deflected upward in spray toward the ceiling.

322. Shows the position of the levers and the fusible joint.

323. DOMESTIC REFRIGERATOR.   Water enters the coil at A, and is drawn into the tank at B, in which a quantity of nitrate of

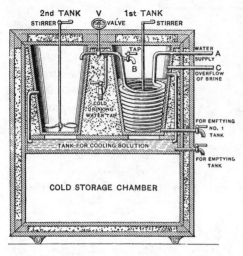

ammonia has been placed equal in weight with the water. The cold produced in the mixture cools the water in the coil, from which drinking water may be drawn. The overflow from the tank siphons to the refrigerator tank below for cooling the storage chamber. For greater refrigeration, or for making a block of ice or freezing a carafe, a quantity of nitrate of ammonia may be placed in the second tank and cold water from the coil drawn to it by the valve V, when the solution temperature falls to zero. The solution or brine from the second tank overflows to the storage tank.

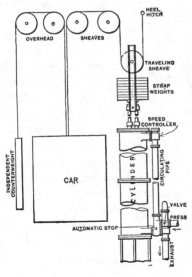

324. COUNTERBALANCING HYDRAULIC ELEVATORS. Showing the arrangement of valves, cylinder, circulating pipe for up and down motion of the hydraulic piston and the distribution of the counterweights to equalize the power. The details of operation are shown by the lettering on the cut.

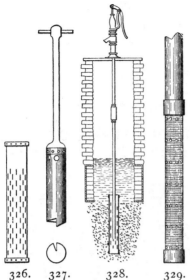

325. RE-ENFORCING WELLS. The re-enforcement of wells in times of drought may be readily made by making a cylinder of galvanized iron, punching it with a thin chisel, as shown in the cut, 326, inserting it and pushing it down in the bottom of the well, 328, and boring out the sand with a sand auger, 327. A drive-strainer tube may also be driven and the sand drawn by an auger. Strainer points are also used and disconnected near the bottom of the well. The supply of water may often be largely increased by these methods.

326.   327.   328.   329.

329. Strainer tube and pipe for direct connection with a pump, as shown in No. 328.

330. SIPHON WATER RAM.   B, a chamber in the apex of a siphon.   C, a flap valve on an arm and spindle extending to outside

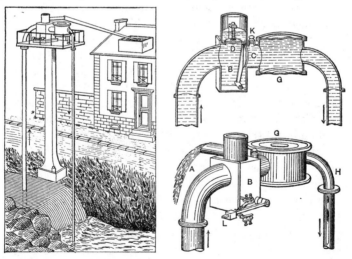

of chamber and held open by the lever and weight L, with its movement adjusted by the springs above and below the lever.   D, discharge

valve.   G, a chamber with elastic heads or diaphragms of thin corrugated metal, for an air chamber and to prevent hammer.   K, plug
for filling the siphon with water or by the suction of an air pump.

Will lift water about 14 feet with a water fall on the siphon legs of
6 feet, and deliver ⅓ of the total supply.

331. Section showing valves and air chamber.

332. Outside view, showing valve lever, weight, and springs.

332A. UNIQUE APPLICATION OF HYDRAULIC POWER.
In a certain river there is a high-water period in the spring when a

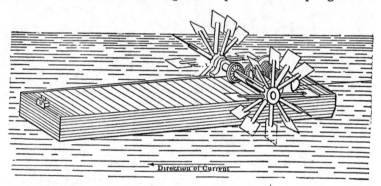

fair sized ocean steamer could easily manœuver.   At the place in
question there is 22 feet fall per mile, which causes a rapid current.
A shipper wanted to transport some material to an island located in
the center of the river at a point where the river was about two-thirds
of a mile wide and the island was about seven-eighths of a mile up
stream.   He had no power except that which nature was producing,
and by means of this and a wire cable about ¼ inch in diameter, one
end of which was attached to a tree on the island, the other fastened
to a reel on board the flat boat, which in turn was connected by gears
to a shaft that carried paddle-wheels on each side of the boat, he
accomplished his task.   After the boat was loaded, the paddle-wheels
were connected to the reel, which was slowly revolved by the action
of the current and commenced to reel up the wire cable, at the same
time drawing the boat and its cargo up to the island.   The accompanying sketch illustrates the scheme used.

SECTION VIII.

AIR-POWER MOTORS AND APPLIANCES.

Section VIII.

AIR-POWER MOTORS AND APPLIANCES.

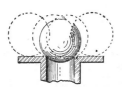

333. PNEUMATIC BALL PUZZLE. A ball laid on the mouth of a flanged tube, as in the cut, can not be blown off by an air jet, but will continue to roll around on the flange, as shown by the dotted lines.

334. PNEUMATIC DISK PUZZLE. A light circular plate with pin guides can only be lifted a small distance by an air jet from the

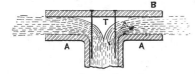

flanged tube. The theory is that the momentum of the air as it suddenly spreads to a larger circumference causes a partial vacuum near the outer edge, thus holding the plates so near together that their circumferential area corresponds with the area of the central jet.

335. PNEUMATIC BALL PUZZLE. A light ball is held in a jet of air from a vertical to an angle of about 30° and revolves with considerable velocity.

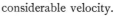

A light ball placed in a conical cup over a jet of air will be held there and not driven off when the cup and jet are reversed.

A card placed on an inverted flanged jet of air, as at V, V, will not drop, even with a considerable weight hanging to it.

336. Inverted nozzle and ball.

337. Inverted nozzle with ball attached to plate.

338. PNEUMATIC FAN. Compressed air is a ready means of operating a fan in shops where it is used for other purposes. By a

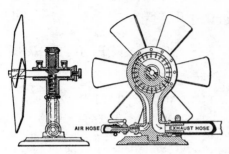

simple air motor, as shown in the cut, with 60 to 80 pounds pressure a high speed may be obtained in the fan which will throw a current of air 25 or more feet, and if the exhaust air mingles with the current its cooling effect will be greatly increased.

339. Cross section, showing motor wheel and pipe connections.

340. THE SIROCCO FAN BLOWER. The particular feature

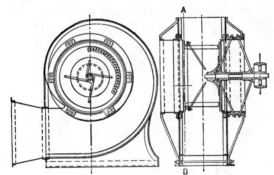

of this fan is in the narrow curved blades set in the periphery of the wheel and close together, which prevents local eddies and greatly increases the efficiency of the fan. Davidson patent.

341. Section showing shaft and bracing to the blade drum.

342. AËRIAL TOP. A small windmill made of any convenient material that is light. A stem at the center to drop into the ring handle, on which the string is wound. The rapid rotation made by quickly pulling the string will force the top to a height of 150 to 200 feet. There should be a ring of brass wire around and fastened to the outside of the fan to give it momentum.

343. Ring handle as held in the hand.

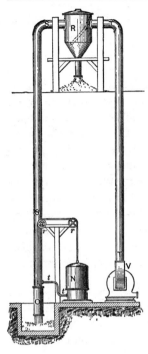

344. PNEUMATIC GRAIN ELE-VATOR. French.

V, a duplex vacuum blower.

T, the vacuum pipe.

R, receiver with a wire screen covering the mouth of the draught pipe.

S, delivery pipe.

O, an air regulator covering the slots in the end of the suction tube S, to regulate the proportion of air and grain entering the tube.

N, a diaphragm chamber to balance the sleeve O, and controlled by the vacuum pressure in the pipe S, through the small tube t, t.

By this pneumatic process of elevating grain, the dust is separated and discharged through the blower and the grain is aërated and dried.

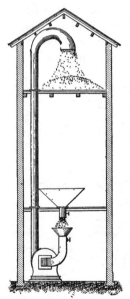

345. PNEUMATIC GRAIN ELE-VATOR. French. A high-speed blower receives the grain from a hopper in a regulated stream through a funnel with sufficient air to make it semifluid, in which condition it passes through the blower and is forced to the desired elevation. By this method grain is dried and aired, and by substituting a chute from a grain bin to the hopper aëration and transfer to other bins is easily made.

A cleaning process by the passage of the grain through the blower, but the dust is deposited in the bin with the grain. Does well for a transfer system only.

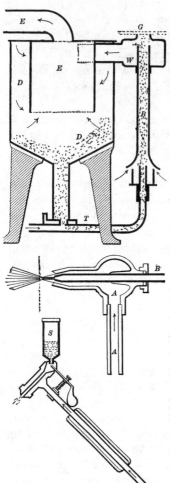

346. SAND-BLAST APPARA-TUS. Exhaust or vacuum type, in which the sand is returned to the supply chamber D, after its use in doing work at G. E, E, exhaust chamber and pipe. Sand dropping from the chamber D is carried along the pipe F by the incoming air to the bell nozzle of the blast pipe B, impinges upon the glass at G, and falls into the chamber W, is drawn into the main chamber D, and falls to the bottom, while the fine dust is carried off in the exhaust. The glass plate at G is moved over the opening of the blast tube for evenly sanding the surface.

347. SAND-BLAST JETS. In the upper figure the sand enters the tube B, by gravity or otherwise, and the air blast through the tube and chamber A, issues in an annular aperture around the sand and compresses the blast to a pencil of abrasion for free-hand pencil work. The lower figure is for the same purpose, but carries a sand box with a regulating valve.

348. Hand sand-blast nozzle with sand reservoir, *s*, for light work.

349. AIR-MOISTENING APPARATUS for textile mills. A jet of high-pressure steam is projected through conical funnels, drawing

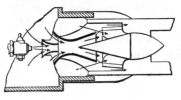

in and mixing air with the steam and spreading the vapor over a cone, and by distributing to various points in a room equalizes the moisture as well as controls its hygrometric intensity.

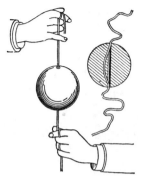

350. MAGIC BALL. A crooked hole is bored through the ball through which the string is passed. By a slight tension of the string, the ball may be stopped or slid down at will of the holder of the string. An amusing trick.

351. Section of the ball showing the crooked hole and string.

352. GYRATING BALLS. This toy consists of two wooden balls of the same diameter connected by a slender elastic rubber band attached by staples.

To prepare the toy for operation, it is only necessary to twist the rubber band by holding one of the balls in the hand and rolling the other round in a circular path upon the floor by giving to the hand a gyratory motion. As soon as the band is twisted, the free ball is grasped in the hand, then both are released at once.

The untwisting of the rubber band causes the balls to roll in opposite directions in a circular path, and centrifugal force causes the balls to fly outwardly. By virtue of the acquired momentum, the balls continue to rotate after the rubber band is untwisted, so that the band is again twisted, but in the opposite direction. As soon as the resistance of the band overcomes the momentum of the balls, the rotation ceases for an instant, when the band again untwisting revolves the balls in the opposite direction, and the operation is repeated until the stored energy is exhausted.

353. MEGASCOPE. A lantern which may have an arc light, a lime light, or a strong lighting lamp, throws its light upon an ob-

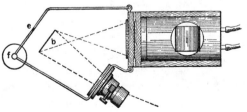

jector picture at *b*, in the focus of a camera lens in a frame or box attached to the lantern as shown, for projecting an enlarged image upon a screen.

354. PNEUMATIC MOISTENING APPARATUS. Compressed air is supplied to the atomizing nozzles at various points in a

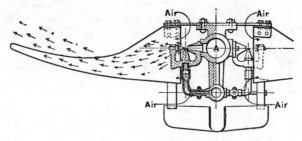

factory by the main pipe A. The water is supplied by the smaller pipe below to the water nozzles, and is atomized and vaporized in contact with the air from the nozzles together with the induced air drawn in by the jets. The wings guide the vapor toward the ceiling, and also collect the excess of water and conveys it to a trough below.

355. THE PANTANEMONE. A stationary windmill in operation in France. Two plane surfaces in the form of semicircles are

mounted at right angles to each other upon a horizontal shaft, and at an angle of 45° with respect to the latter. It results from this that the apparatus will operate (even without being set) whatever be the direction of the wind, except when it blows perpendicularly upon the axle, thus permitting (owing to the impossibility of reducing the surfaces) of threescore days more work per year being obtained than can be with other mills so claimed in proportion to the work of the old Holland mills.

356. A KANSAS WINDMILL. Made with canvas sails with the axle set on the meridian so that it runs with any wind with north or south in it. It is crude and homemade. Every farmer can make one for pumping water, churning, and many small wants for power.

357. SAILING WAGON. Across the wide forward end of the triangular frame extends an axle to which wheels are journaled. The

short axle of the rear wheels is pivoted by a kingbolt to the narrow end of the frame. To the short axle is attached a gear wheel into which meshes a smaller wheel secured to the lower end of a vertical shaft journaled in bearings fastened to the frame. Upon the upper end of this shaft is a handwheel or tiller, by means of which the wagon may be guided. The speed of the wagon is regulated by brakes upon the front wheels, connected with an upright lever pivoted in the middle part of the frame and provided at its upper end with a crosshead, so that it can be operated either with the hands or feet. A mast fastened to the middle forward part of the frame is provided with a sail and appliances for raising, lowering, and controlling the sail in the same manner as an ordinary sailboat.

358. SAIL-RIGGED MERRY-GO-ROUND, St. Malo, France.

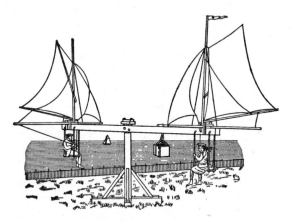

A swinging beam on an anchored post balanced by a movable box of sand.

Each end of the beam has a crossbar on which is rigged a mast with mainsail and jib.

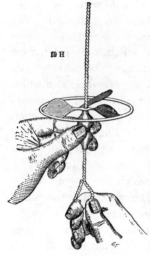

359. FLYING PROPELLER. At the center of the wheel there is a square hole in which is loosely fitted a twisted square rod, and upon this rod, below the wheel, is placed a wooden sleeve, the bore of which is large enough to allow the rod to be readily drawn through it.

The wheel having been placed upon the rod -as shown in the engraving—the wooden sleeve is grasped between the thumb and finger of one hand, the eye at the lower end of the rod is grasped by the other hand, and the rod is drawn quickly downward, thus imparting to the wheel a very rapid rotary motion which causes it to rise to a great height in the air as it leaves the rod.

360. A KITE WITHOUT A TAIL. All the calculations necessary in order to obtain the different proportions are based upon the

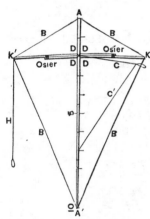

length of the stick, A'A, employed. Such length being found, we divide it by ten, and thus obtain what is called the unit of length. With such unit it is very easy to obtain all the proportions. The bow, K'K, consists of two pieces of osier each five and a half units in length, that form, through their union, or lap, a total length of seven units.

After the bow has been constructed according to these measurements, it only remains to fix it to the stick in such a way that it shall be two units distant from the upper end of the stick. The balance, or belly band, CC', whose accuracy contributes much to the stability of the whole in the air, consists of a string fixed at one end to the junction, D, of the bow and stick, and at the other to the stick itself at a distance of three units from the lower extremity. Next, a cord, B, is passed around the frame, and the whole is covered with thin paper.

Before raising the kite, the string, which hangs from K', is made fast at K in such a way as to cause the bow to curve backward. This curvature is increased or diminished according to the force of the wind.

Nothing remains to be done but to attach the cord to the balance, or belly band, and raise the kite.

361 THE EDDY TAILLESS KITE. The sticks should be made of clear spruce, as this has been found to be less liable to bend

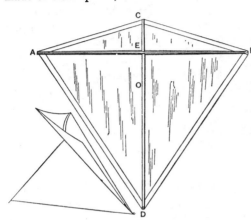

under strain or break at the cross stick.

Cross section of each stick is $\frac{5}{16}$ by $\frac{1}{2}$ inch.

Kite stick A B $=$ $68\frac{4}{10}$ inches.

Kite stick C D $=$ 60 inches.

O $=$ center of gravity, which is 35 per cent of C D from the top of C D.

C E $=$ 18 per cent of C D in both strong and light wind kites. The thin manila paper should be put on the kite slightly loose. The deepest part of the bow of the cross stick A B should be about $\frac{1}{10}$ of the length of A B. In bending A B great care is required to see that the bend on each side of the point of junction at E is equal. The slight bagging inward of the paper covering triangles A E D and B E D should be equal. If the kite flies sideways, owing to inequality, it can be partly remedied by tying small half or quarter ounce weights at A or B. The hangers or belly band drawn in the side view of the kite, fastened to E and D only, make a right angle at E and an acute angle at D.

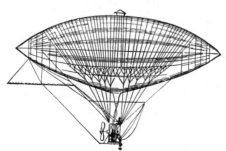

362. TISSANDIER'S ELECTRIC AIR SHIP. Paris, 1883. This air ship attained a velocity of eight miles per hour, operated by an electric motor with current from a storage battery.

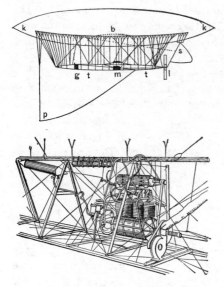

363. SANTOS-DUMONT AIR SHIP. Showing the framework and its attachment to the balloon, the position of the propeller, rudder, and gasoline motor, which is in the center of the framework and balloon to balance the ship.

364. The motor has four cylinders, air cooled, for which purpose a fan blower is operated by the motor.

365. GIFFARD'S STEAM-PROPELLED AIR SHIP. One of the earliest of the present type of air ships. This aërial steamer ascended from the Hippodrome in Paris, September 25, 1852, to a height of 5,000 feet. After a successful sail, landed safely.

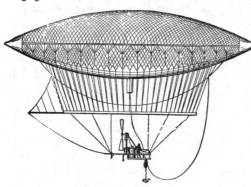

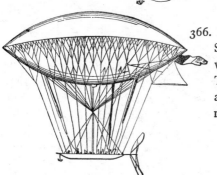

366. **DUPUY DE LOME'S AIR SHIP.** It carried twelve men who turned the propeller. This air ship ascended in 1872 and attained a speed of six miles per hour.

367. THE CAMPBELL AIR SHIP. The propelling power was by cranks operated by the aëronaut, one on the lifting propeller and

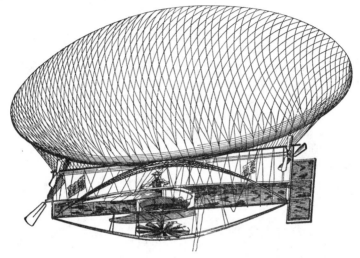

one for the driving propellers. The operator had good control in the trials made at Coney Island, N. Y., but the ship was finally blown to sea and lost, 1889.

368. POWER FLYING MACHINE. Maxim's type. The application of power to flying machines has been several times success-

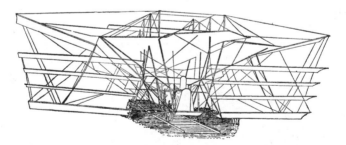

fully tried. Professor Langley's aërodrome, which resembles an enormous bird of steel, was tried with much success in May, 1896. It rose easily and soared in the air in large spiral curves of 100 yards diameter, reaching a height of about one hundred feet and moving about half a mile. The steam then gave out and the propeller stopped, but the machine, instead of tumbling to the earth, settled slowly and gracefully downward and reached the surface without damage. Its greatest speed was nearly at the rate of twenty miles an hour. Maxim's experiments are still more interesting. He constructed a flying machine on a large scale, its total weight when loaded being 8,000 pounds, this including engines, boiler, fuel, stores, and three persons. The boat-like body was moved by a powerful propeller, and the lifting mechanism consisted of a great aëroplane, with smaller ones projecting like wings, the extreme width being 105 feet, length 104 feet, total area 5,400 square feet. He had constructed a railway along which this machine moved on wheels, the pressure on the rails decreasing as the speed increased. In a notable experiment, made in June, 1894, the whole machine was lifted for a brief interval from the ground.

369. RENARD & KREBS ELECTRIC AIR SHIP.   Paris, 1884. The electric motor was operated by current from storage bat-

teries. The form was peculiar, being somewhat like a fish, with the propeller at the head. It was claimed to have attained a speed of twelve miles per hour.

370. GRAIN-DRYING APPARATUS. For tumbling grain or other material in an inclined cylinder with a blast of warm air. A, a brick

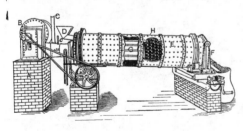

box in which coke is burned, or a flue to convey waste heat from any furnace. B, compound wrought-iron fan, which will draw waste heat from a distance of 50–100 feet. C, chimney and valve, to carry off smoke when fire is first lighted. *c*, thermometer or pyrometer. D, feed hopper, into which the grain is conveyed by an elevator from below, or by a chute from an upper floor. E, cylinder. F, elevating gear for raising and depressing cylinder.   G, air duct, made of different sec-

tions to suit different products.   H, part of the outer shell removed to
show the cells in which the grain is carried up and poured out in a continual stream; the number and pitch of these cells is also varied for
various products.

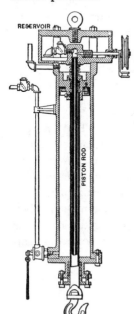

371. PNEUMATIC LIFT.   Ridgway
type, oil governed.   A tube extends from a
reservoir down the inside of the hollow piston
rod.   The reservoir is filled with oil.   When
the hook is lowered, the oil is drawn into the
piston rod through the check valve.   The
chain wheel and needle valve govern the flow
of oil, which by its non-compressibility prevents vibration of the load and holds it at
any desired height.   This device eliminates
the jerky motion of the plain air lift.

Type of the Craig Ridgway & Son Company, Coatesville, Pa.

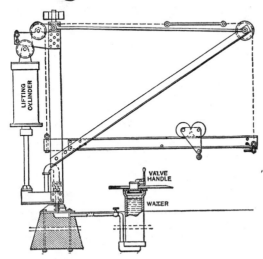

372. AIR-OPER-
ATED HYDRAU-
LIC CRANE. Com-
pressed air pressure in
a supplementary cyl-
inder forces the water
into the lifting cylin-
der.   A water valve
governs the flow of
water and holds the
weight steady or
locked by closing the
valve.

373. VALVE-LIGHT VENTI-LATOR. The valves are of glass set in frames and hung to swing, controlled by connecting rods to a vertical pole extending down within reach of the hand.

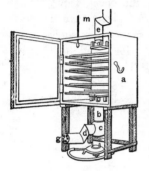

374. FRUIT-DRYING APPARATUS. A box, *a*, arranged for receiving perforated shelves or netting on frames. A fresh-air inlet, *g*, and heating chamber, *c*, under which a lamp is placed. A deflecting plate, *h*, to spread the warm air evenly through the box and another at the top for gathering the air to the ventilator *e ; m*, a thermometer. Temperature should be 100° Fah.

SECTION IX.

GAS AND AIR-GAS DEVICES, ETC.

Section IX.

GAS AND AIR-GAS DEVICES, ETC.

375. KEROSENE PORTABLE FORGE. French. The apparatus consists of a copper reservoir, P, containing the petroleum,  and traversed by a pump, C, which serves to establish a pressure of air at the surface of the liquid. Above the reservoir, and separated therefrom by a horizontal disk, D, forming a screen to prevent the heating of the reservoir, is placed the stove, so called. In the latter, the kerosene is burned after being vaporized by its passage through a worm, S, heated by the flame. This worm is formed of an iron tube starting from the bottom of the reservoir and ending in a central jet at the other extremity. Upon the tube is placed a cock, B, for regulating the discharge of the oil, and, consequently, the intensity of the flame. Beneath the worm there is an iron cup which is opened at E, and into which, for lighting, is poured a spoonful of amylic alcohol, after care has been taken to fill the reservoir, P, with oil after unscrewing the plug A. The alcohol is lighted, and as soon as the worm is hot the cock is opened, the jet takes fire, and the apparatus is ready for use. Upon the stove there may be placed either a cast-iron pot in which to melt lead or tin, or the tools that it is desired to heat or temper, or the iron tubes to be bent, etc.

376. Section of forge with fire tiled cap for deflecting the heat downward on to the work.

377. PRODUCER GAS GENERATOR. German type. A, door for feeding coke to the furnace B and for blowing up. C, fire-

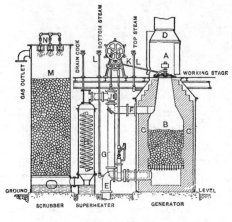

brick walls of the furnace. E, air inlet for heating the furnace of the generator. F and G, gas blow-off pipe, interchangeable to reverse the gas blow. J, valve that automatically closes when A is opened. L, L, steam pipes for alternating the steam blow. H, super-heating coil for heating the steam by the hot gases passing to the scrubber M. N, sprinkler. K, wheel and drum for simultaneously opening and closing the valves J and G and the blast door A.

378. MOND GAS PLANT. Dr. Mond's process, briefly de-scribed, is as follows : The cheapest bituminous slack obtainable is

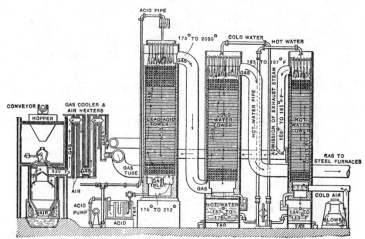

mechanically deposited in hoppers above the producers. From this it is discharged into the producer bell, where the heating of the slack takes place, and the products of distillation pass down into the hot zone of fuel before joining the bulk of the gas leaving the producer.

The hot zone destroys the tar and converts it into a fixed gas, and prepares the slack for descent into the body of the producer, where it is acted upon by an air blast which has been saturated with moisture and water superheated before contact with the fuel. The hot gas and undecomposed steam leaving the producer pass first through a tubular regenerator in the opposite direction to the incoming blast. An exchange of heat takes place, and the blast is still further heated by passing down the annular space between the two shells of the producer on its way to the fire grate ; then the hot products from the producer are further passed through a " washer," which is a large, rectangular, wrought-iron chamber with side lutes ; and here they meet a water spray thrown up by revolving dashers, which have blades skimming up the surface of the water contained in the washer. The intimate contact thus secured causes the steam and gas to be cooled down to about 194° F., and by the formation of more steam tending to saturate the gas with water vapor at this temperature, then passing upward through a lead-lined tower, filled with tile to present a large surface, the producer gas meets a downward flow of acid liquor, circulated by pumps, containing sulphate of ammonia with about 4 per cent excess of free sulphuric acid.

Combination of the ammonia of the gas with the free acid takes place, giving still more sulphate of ammonia, so that to make the process continuous, some sulphate liquor is constantly withdrawn from circulation and evaporated to yield solid sulphate of ammonia, and some free acid is constantly added to the liquor circulating through the tower. The gas, being now freed of its ammonia, is conducted into a gas-cooling tower, where it meets a downward flow of cold water, thus further cooling and cleaning it before it passes to the various furnaces and gas engines in which it is used.

379. AIR AND VAPOR GAS GENERATOR. A rotary air pump driven by a weight forces air through a gasoline carburetor,

which becomes saturated with vapor and distributed for illumination. The internal arrangement of the carburetor may be of any design that will expose a large surface of the gasoline to the air.

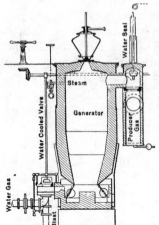

380. WATER-GAS PLANT. Lowe type. An iron cylinder lined with fire-brick. Air is blown in at the bottom for heating the coal or coke. Then steam is blown in at the top, passing through the hot fuel and discharged at the bottom as water gas. Fuel is fed through the hopper at the top. By reversing the blowing by steam and air, producer gas is made and discharged through the side pipe at the right.

381. THE "WELLS LIGHT." The light is produced by passing kerosene oil through a heated burner, where it is generated into gas,

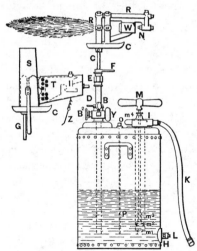

the gas burning in a large, powerful flame which needs no protection, and will stand any weather.

The oil is forced into the tank by the pump M, through the hose pipe K, until it is two-thirds full, compressing the air already in the tank to about 25 lbs. pressure.

The burner is heated by burning a little oil in the dish C, the heat being concentrated around the burner tubes by the chimney S. In seven or eight minutes the burner will be sufficiently heated; the valve B^2 is then opened a little and the oil from the tank is forced by the air pressure into the heated burner, where it is converted into gas, which issues from the jet N, mixing with sufficient air in the cone W, where it may be ignited; the chimney is then removed, and the flame passing through the rings of the burner, maintains the heat and gives a clear, white light, free from smoke or spray. A few strokes of the pump every few hours is all that is required to renew the pressure —and oil or air can be pumped into the tank while the light is burning.

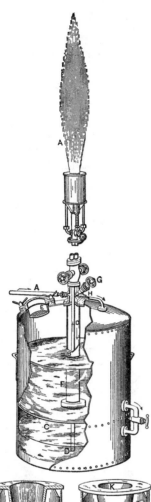

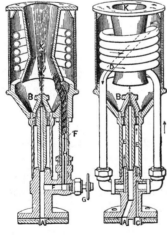

382. LUCIGEN LIGHT. For outdoor lighting. The lucigen employs the most diverse oils—crude and rectified petroleum, naphthas, oil of tar, vegetable oils, waste lubricating oil, etc.

The oil is poured into the reservoir through the sieve, E, which retains the solid particles, if there are any. It collects in a compartment, F, which communicates with the lower part, D, through a tube provided with a cock shown to the right of the engraving. The compressed air enters through the pipe, A, descends through the tube, B, into the air chamber, C, and causes the oil to ascend in the tube, D, which leads to the burner. The oil reservoir has a double bottom that forms a feed chamber that can be filled during the operation of the system.

Sections 383, 384 will allow the operation of the burner to be understood. The oil enters the tube, A, under pressure, and makes its exit through a cylindrico-conic ajutage placed within the lamp. This ajutage is capped by a second ajutage, B, serving for the passage of the air and the atomized oil. The air enters through a conduit, C, parallel with the tube that conveys the oil, and is heated by passing through the coil and further heats the oil in its annular passage, E, to the atomizing burner.

159

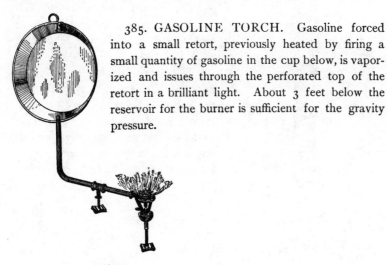

385. GASOLINE TORCH. Gasoline forced into a small retort, previously heated by firing a small quantity of gasoline in the cup below, is vaporized and issues through the perforated top of the retort in a brilliant light. About 3 feet below the reservoir for the burner is sufficient for the gravity pressure.

386. KEROSENE SOLDERING FURNACE. Air is compressed in the oil tank by the rubber bulb by which the oil under pressure is forced through a needle valve to a vaporizing retort and by a Bunsen-burner jet is mixed with air and forced into the burner tube, which is perforated with small holes that feed the heating flame.

387. KEROSENE OIL BURNER for stoves. The coil of iron pipe in the box is the vaporizer, terminating in the cross pipe and two

jet burners. The cones deflect the heat upward, allowing sufficient heat to the coils for vaporizing the oil. For starting the burner, asbestos mats wet with oil are placed under the vaporizing coils and fired.

388. KEROSENE COOK STOVE. National Oil Heating Co.
type. Air is compressed in the oil tank above the oil, which drives the

oil to a vaporizer in the burner pan, where the oil is vaporized and forced through a combination air jet to the chamber of the burner box, which is filled with small tubes which supply additional air for complete combustion. See Figs. 386 and 387 for similar burners.

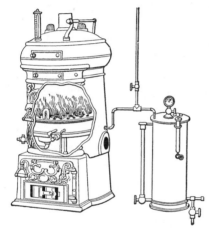

389. KEROSENE HEAT-ER. National Oil Heating Co. type. Air is compressed in the oil tank above the oil, which drives the oil to a vaporizer in the burner pan, where the oil is vaporized and forces a combination air jet to the chamber of the burner box, which is filled with small tubes which supply additional air for complete combustion. See Figs. 382, 383 and 384 for similar burners.

390. GAS GRAVITY BALANCE. A glass globe is nicely balanced on a hollow beam with a pointer at the opposite end and a scale.

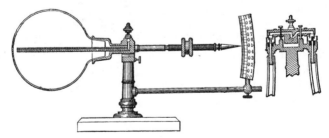

The inlet and outlet for the gas at the knife-edged pivots are sealed in mercury cups so as to make a free passage for the gas without affecting the balance, as shown in the section.

391. GAS-FIRED LIMEKILNS. The illustration shows a double kiln, but the two parts are independent of each other, and may

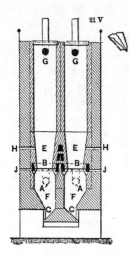

be worked separately. The gas from the producers enters the kiln at A, the flow being regulated by valves at B. At C are doors by which the air necessary for combustion enters, the air and gas meeting at B B. The lime is burned in the chambers, E, and is afterward cooled as it descends in the zones, F, by the air passing in at the lower part. The waste heat is conducted away in the upper part of the kiln through the chimney openings at G. At H are sight holes for judging the heat of the kiln, and J are holes to admit air when the flues have to be burned out. The fuel used in the gas producers is ordinary slack. A special feature is the method of constructing the central partition wall, this having air-cooling and circulating cavities as shown. The lime produced is free from clinkers.

391A. SNOW REMOVER. This apparatus is provided with a hydrocarbon burner and a tank for containing suitable hydrocarbon liquid, which is fed through a pipe to the burner tube, which is directed downwardly on the pavement so as to come in contact

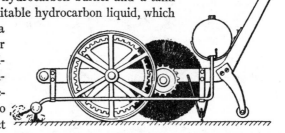

with the ice and snow. By this means the ice is melted and the large brush which revolves at a high speed readily removes the melted ice. Back of the brush is placed a scraper, which comes in direct contact with the pavement, thus removing all ice left in the passage of the brush.

SECTION X.

ELECTRIC POWER AND DEVICES.

Section X.

ELECTRIC POWER AND DEVICES.

392. ELECTRIC CABLE-MAKING MACHINE. A revolving frame carrying the required number of wire bobbins. The strands are gathered by passing through a die and covered in reverse wrappings

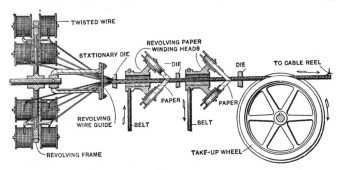

by passing through revolving heads mounted with paper spools, and through dies to compact the paper winding. The cable is drawn by the take-up wheel, which is conical on the face and draws by several frictional windings.

393. CHLORIDE ACCUMULATOR or storage battery. The alternate perforated plates are filled with peroxide of lead and spongy lead. The principal action when the cell is charged is the formation of lead peroxide on the positive plate and spongy lead on the negative. When the cell is discharging, the lead peroxide gradually changes to lead sulphate and the metallic lead on the negative plate also changes to lead sulphate.

c, b, recesses in the lead plates for receiving the spongy lead. Sulphuric acid 1 part and water 8 parts for filling the cell.

394. ELECTRIC WIRE INSULATING DEVICE. Four-spool system. First layer of silk wound left hand, second layer of white

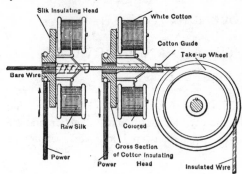

and colored cotton wound right hand. The two pairs of spools and frames revolve in opposite directions that the wrappings may cross each other. The take-up wheel regulates the traverse of the wire through the machine.

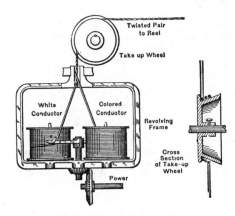

395. ELECTRIC WIRE DOUBLING DEVICE. The previously covered wires, one white and one colored red to distinguish them in wiring, are wound together and drawn over the conical take-up wheel, as shown in the right-hand section, with several friction turns to regulate the twist rate.

396. ELECTRIC WIRE INSULATING DEVICE. Braiding system, in which a variegated color is given by using different colored

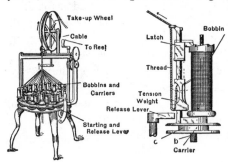

thread spools or bobbins. The right-hand figure shows the details of the bobbin latch or let-off and the tension weight carrying the thread. The grooved disk c rides in the traverse slots and is carried along by the pin b.

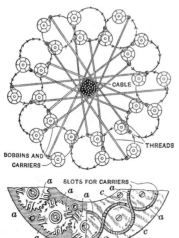

397. CABLE COVER BRAID-ING MACHINE. Details of the bobbin motion, bobbin carrier gear, and slotted guide plate. *a, a* are the guide fingers on the gears that push the bobbins along the grooves, which by their crossing grooves carry the spools out and inside each other.

398. Shows the gears beneath the slot-plate, each with its four guide fingers that mesh to carry the bobbin slide into the opposite slot at each quarter revolution.

399. WIRE-COVERING MACHINE. G is the wire reel from which the wire is drawn through the machine by the geared rollers E, E,

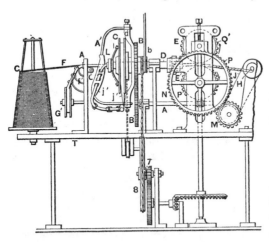

and wound upon a drum at H. A rubber ribbon is wrapped upon the wire through the guides and eye at A. The spool C delivers the wrapping ribbon through the eyes of the revolving yoke J, J. The gears B, B drive the shaft A² and worm *a*, giving motion to the drawing rolls E, E, and the winding drum H, through the gears N, M. *b* is the driving belt. The chain and gear below the bed of the machine are for change motion to the wire feed by a train of spur and bevel gear.

400. SHUNT-WOUND DYNAMO. The brushes B, B are connected to the main lines, M, M, supplying the outside circuit. The

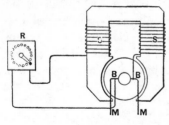

field magnet coils *s, s* are connected
in shunt across the armature at the
brushes B, B. The field coils are of
fine wire and many turns, with a resistance of many times that of the
armature, in which is interposed the
resistance box R.

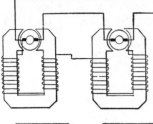

401. SHUNT DYNAMOS connected in series. The shunt winding is
connected across both dynamos and
the other end of the winding to the
opposite poles in the armature brushes.

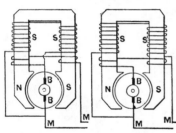

**402. SHORT AND LONG
SHUNT.** Compound dynamo winding. Shows the two ways of making
the terminal connections of the windings. Right-hand figure is short
shunt.

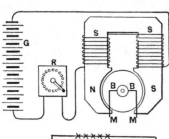

**403. SEPARATELY EXCITED
DYNAMO.** B, B are the brushes
of the armature circuit to the lines
M, M. The battery G supplies current to the field winding only, which
is regulated by the resistance box R.

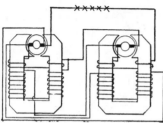

**404. COMPOUND WOUND
DYNAMOS** in series. Shunt coils of
each excited from the other dynamo.
The series coils are in series with the
main circuit.

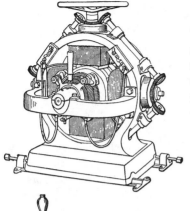

405. MULTI-SPEED ELEC-TRIC MOTOR. A handwheel and set of bevel gears draw the field magnets away from the armature for varying the speed by changing their distance apart. Model of the Stow Mfg. Co.

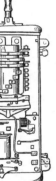

406. DRUM CONTROLLER in which variations in speed are controlled by throwing into the circuit resistance in sections suitable to the requirements for different speeds.

The details of construction vary greatly to meet the purpose for which they are to be used, and are made in rheostatic or resistance type, or in series parallel type for shunting or short-circuiting one of the motors.

407. COMMUTATOR CONSTRUCTION. Edison type. A cast-iron or brass sleeve s is bored to fit the shaft. On the back end

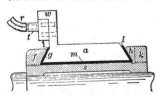

is secured a steel collar f, which is coned at g to fit the conical ends of the bars, a conical steel ring h slips up against the other end, and the bars are clamped up by means of the steel nut k. The insulation is entirely of mica, as this is the only insulating material that has been found suitable for the insulation of commutators. The sleeve s is insulated by a cylindrical body insulation m, against which the bars are clamped. The end insulations i and l are of mica built up into conical form and pressed into shape in suitable molds. With this particular style of commutator the leads r from the armature winding are soldered into an ear or cup t which is screwed to the ear w on the commutator bar a by means of flat-headed, countersunk screws.

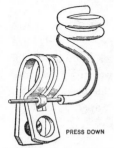

408. SPRING BINDING POST. A quick method of changing electric-wire connections—press the spring and push in the wire.

A most convenient binding post.

PRESS DOWN

409. ELECTRIC TRANSFORMER. Used only with alternating currents. The principles of action are in the change of a high electro-motive force or voltage, to a low one, and *vice versa*.

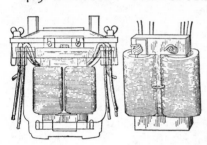

The secondary or low voltage winding is of coarse wire wound next to the soft iron core, with the primary, high voltage, fine wire wound on the outside; thoroughly insulated and provided with means for cooling by air circulation or an oil bath.

410. Shows the form of the core and winding.

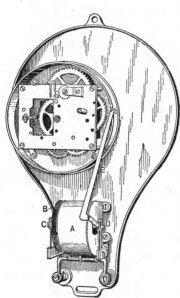

411. RECORDING AMPÈRE METER. Bristol's type. A is a stationary coil or solenoid through which current passes. B is a very thin disk armature of iron secured to a non-magnetic shaft which extends through center of the solenoid A, and is supported at its opposite ends on steel knife-edge spring supports C and D. The recording pen arm E is secured directly to the steel spring support D, and partakes of its angular motion as the armature is attracted to the coil or solenoid by a current passing through the solenoid. The face or recording dial is not shown as it covers the clockwork that drives the dial.

412. NOVEL ARC LAMP. Carbons are held in inclined troughs 33, 19, supported on springs 31, 16, by similar troughs 35, 21, which 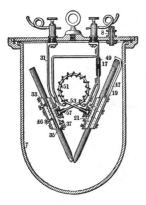 are carried by headed pins 37, and are attached by insulating sockets 57 to the ends of an expansible metal strip 53. The strip is surrounded by a coiled heating resistance 51, connected in series with the carbons, and is thus heated and expanded so as to press the movable troughs 35, 21 against the carbons, and move the carbons apart, when sufficient current is supplied, but to allow the carbons to move together and slip down when the current decreases. Each carbon may be pressed down by a spring 49 placed between its coned upper end and an arm 47 of its holder ; with continuous current, movement of the negative carbon is retarded by screws 46. The expansion strip 53 and heater 51 may be replaced by toggle links connecting the sockets 57 with an arm of an iron core, movable vertically into a series solenoid on the top plate 1. The springs may be steel strips coated with copper. The ends of the heater or solenoid wires are connected to the upper and lower parts of the spring, which is divided by an insulator 17. The lamp may be inclosed by a globe 7, secured by a packed ring and screws to the top plate. A nozzle 8 is provided, through which air may be exhausted from the globe, and another gas introduced.

413. SEARCH–LIGHT MIRROR. Silvered back. The lens 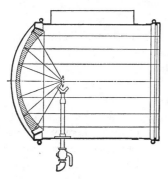 mirror is accurately ground and polished to the exact curvature required to give a perfectly parallel beam, and with a half-foot acetylene flame it will show up the whole road for over 1,500 feet ; the same size flame with the best metal reflector will show only a hundred feet or so.

The unequal curves of the mirror are for the purpose of eliminating the spherical aberration.

414. ELECTRIC ENGINE STOP. Monarch type. The magnets, A, are placed in circuit with an electric battery, and when circuit is

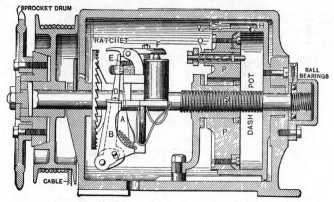

closed by pressing a button, the armature end of the lever B is pulled down, releasing the upper end of the vertical lever D, which also serves

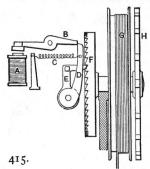

as a *hammer*, striking the lug on the pawl E, throwing it out of engagement with the ratchet, thus allowing the shaft of the stop to revolve and close the valve by means of the sprocket chain attached to the sprocket wheel of the stop, engaging a similar sprocket wheel attached to the throttle-valve stem, the weight on the cable furnishing the power. At the opposite or right end of the stop is a dashpot, which consists of a cylinder

415.

into which the piston P fits closely. On this end of the shaft is cut a square-threaded screw, S, passing through a nut fastened in the center of the piston, P, so that as the shaft revolves, by means of this screw the piston is carried into the cylinder, and the air behind its inner face is compressed, thus forming a complete cushion. The speed at which the stop acts may be very accurately adjusted by turning the by-pass valve V, which governs the amount of air that is forced through the air passage H, as the piston P moves in. Below this by-pass valve V, and in the piston P, is located a releasing valve, O, which can be adjusted at will to open by contact against the bottom of the dashpot, when the throttle valve is near its seat, allowing the compressed air to escape quickly, *after the piston has cushioned*, thus allowing the valve to start again and take its seat softly, but with sufficient force to close it tightly.

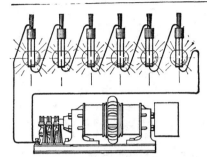

416. SERIES ARC LIGHT-ING CIRCUIT. A multiseries arc dynamo of the Brush system. The circuits may be combined or single, controlled by a switch-board and three part armature and commutator.

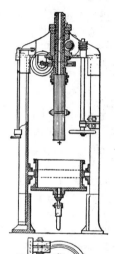

417. ROTATING ELECTRIC FUR-NACE. French design. The upper electrode is swung in two directions and at the same time revolves to cover the entire bed in the pot, also depressed or elevated by the handwheel and gear. The upper end of the electrode bar is round, with a toothed rack. The crucible is charged from a trough and emptied by turn over on its trunnions. The carbon lining of the pot is the negative electrode, not shown in the engraving. The plan shows the worm gear for turning the swinging electrode.

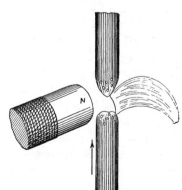

418. ELECTRIC BLOWPIPE. A strong electro-magnet repels the electric arc with such force that it may be used as a blowpipe of high temperature.

A curious example of the repel-lant action of the magnet upon an electric arc.

419. ELECTRIC FURNACE for making calcium carbide. English. The furnace consists of a firebrick casing A, with a magnesia

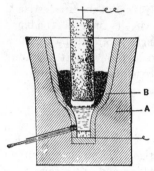

lining B. The shape is conical, and at the bottom the furnace is contracted to form a hearth for the fused carbide. The tapping hole is at the bottom of this contracted part. The lower electrode is a carbon plate, and the upper electrode a massive carbon rod of circular section. The raw material is fed into the annular space between the upper electrode and the magnesia lining in sufficient quantity to inclose and smother the zone of highest temperature.

420. TANDEM WORM-GEAR ELECTRIC ELEVATOR. Siegel-Cooper store, N. Y. Hindley type. Geared 46 to 1 for a speed

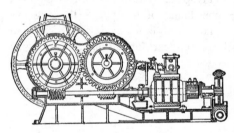

of the car of 100 feet per minute with a motor speed of 470 revolutions per minute. Efficiency from current to car service 70 per cent. The double worm and interlocking gears contribute to the safety of the elevator service.

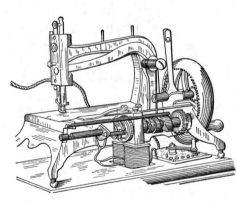

421. ELECTRICALLY DRIVEN SEWING MACHINE. The armature is on the shaft that operates the needle bar and shuttle with a rheostat to control the speed. A pinion on the driving shaft meshing in an internal toothed wheel with handle enables management of the sewing machine by hand.

422. ELECTRIC MOTOR WORM-DRIVEN PUMP. It con-
sists of the motor E, mounted on a base with a duplex pump, which

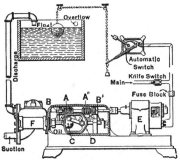

drives the latter by means of double
worm gearing and cranks. The combination of right- and left-hand worms,
A and A′, drives two worm wheels,
B and B′, which mesh together and
thereby balance the thrust of the
worms. The cranks are mounted
on the shaft of the worm wheel B.
One-half the power of the motor
(less friction) is transmitted to the
worm wheel B through the worm
meshing with it; the other one-half is transmitted to it by the meshing
worm wheel B′. By the use of the yoked extended piston rod D and
short connecting rod the combination is made unusually compact.

The float in the tank is the governor of the pump, through the
automatic switch.

423. ELECTRIC INCUBATOR. German. A basket filled with
hay or fine straw upon which the eggs are laid. The cover consists of

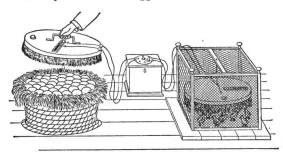

a layer of soft down attached to a circular box containing coils of wire.
The latter are heated by an electric current whose temperature is regulated by a thermometer placed on the cover. When the heat becomes too
great, the rise of the mercury cuts the coils out of circuit and allows them
to cool. A coop for the chicks, in which the cover can be raised to accommodate with their growth. The only attention required is to sprinkle
the eggs with fresh water and to turn them once a day. A rheostat
regulates the current for a nearly uniform temperature of the heating
coils.

424. ELECTRICAL SOLDERING COPPER. The resistance or heating coil is composed of small iron wire wound on insulating

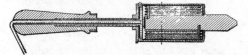

material (asbestos cloth). The coils are wound far enough apart to prevent short circuiting and their electric connections insulated and carried through the handle.

425. ELECTRIC WELDING APPARATUS. A shunt dynamo charges the 50 accumulator cells in series ; a voltmeter and an ampère-

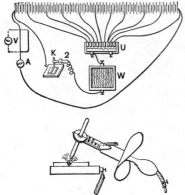

meter are inserted at V and A. From the positive terminal of every fifth cell a wire leads to a plug switch board U; from U the current passes through a variable resistance W, and from thence through a flexible cable to the carbon holder 2 and the carbon pencil K. The operator manipulates his holder 2, the metal to be fused, placed upon the table, being joined directly to the negative terminal of the battery. By inserting the plug in the switch board U, the operator may obtain currents from 5 cells, twice 5, and so on to 10 times 5 cells.

426. Carbon holder, carbon pencil at work.

427. ELECTRIC WELDING. The operator wears strong leather

gloves, and his hand is further protected by a metal screen fixed on the holder. He looks at his work through a dark glass, which protects both his eyes and face from the radiated light and heat better than ordinary dark spectacles would do. The lungs also may need protection from the vapors of copper, lead, and other metals or alloys. When possible, means should be provided to carry off such vapors with a blast of air. The construction of the holder permits of a quick replacement of the carbon pencil. See Fig. 425 for details.

428. ELECTRIC REVOLV-ING CRANE. 150 tons hoisting capacity. Erected by the Newport News Ship Building & Dry-dock Co. A lifting and revolving crane that has been built on the most modern principles in mechanical construction for compactness and efficiency.

429. ELECTRO-MAGNETIC TRACK BRAKE. The track-brake shoe is placed between the two pairs of wheels, and, instead of

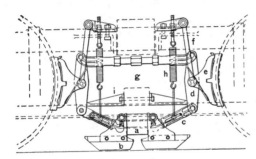

being forced upon the rails through an effort from the car, is drawn to the rails by an electro-magnet suspended from the car, thereby not merely adding its friction to the unimpaired friction of the wheel brake but also actually increasing the rail pres-sure of the wheels to the extent that the supporting springs for the track shoes and magnets are in tension through the descent of the track shoes to the rails. The electro-magnet a, dividing the track-brake shoe b into two parts, is secured by pins to the two push rods c, and sus-pended at a proper distance above the rails by the adjustable springs h. The push rods are secured by pins to the lower ends of the brake levers d, which are connected at their upper ends by the adjustable rod g and are pivoted at an intermediate point to the brake-shoe holders e, carry-ing the wheel-brake shoes, and the hanger links f, suspended from the truck frame. The push rods c are telescopic, as shown in the sectional view of the one at the left, so that a movement of the track shoe toward the right, relative to the truck frame, causes the wheel-brake shoe at the right to be applied to the wheel and the connection *g* to be moved to the left, thereby applying the wheel-brake shoe at the left, the stop *i* preventing the lower end of the brake lever at the left from following the track-brake shoe.

430. ELECTRO - MAGNETIC CLUTCH. Reverse change speed. A is the motor, of which the armature shaft is extended at both ends to receive pinions B and C.

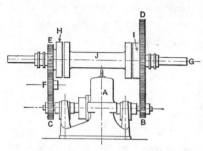

Pinion B drives gear D, and pinion C drives gear E through idler F. B is smaller than C, and D is larger than E. It follows that gear D runs slower than gear E, and in the opposite direction. Both gears D and E run loose on shaft G, and each of them is keyed or bolted to a part H or I of the magnetic clutch, which parts are iron-clad electro-magnets that can be energized or de-energized at will. J is the armature or keeper, which is keyed to shaft G, but can slide over it. If I is energized, J is attracted toward it and is compelled to revolve with gear D, thus giving the driving shaft a slow motion. If H is energized, J is attracted toward it and follows the motion of gear E, thus giving the driving shaft a fast motion in the opposite direction. The shifting mechanism is so arranged that only one electro-magnet is in action at one time.

431. ELECTRO-MAGNETIC CLUTCH. The figure shows a magnetic clutch with its armature and shaft in cross section. A and A

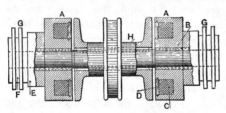

are the magnets, each provided with a brass bush B. The coil of wire C is placed in an annular groove in the magnet, and held in position by a ring of lead D calked into a recess of the groove. An extension E of the magnet is turned down so as to make a gear fit, and a further extension F takes the collector rings G. As will be seen, there are two collector rings for each magnet : one for leading the current into the coil, and one for the return. H, armature keyed to slide on shaft and may have a belt pulley.

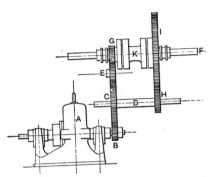

432. ELECTRO-MAGNETIC CLUTCH. Reverse change speed from a single pinion. A is the motor and B the motor pinion, driving a gear C which is keyed to a shaft D. From shaft D the motion can be transmitted to shaft F, either through gears C E and G, or through gears H and I—the clutch arrangement being the same as in Figs. 430–431.

433. WIRELESS TELEGRAPHY. Marconi receiver. j^3 is the coherer tube, with its silver pole pieces, j^1, j^2. The coherer forms part

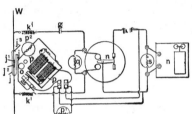

of a circuit containing a local cell, g, and a sensitive telegraphy relay. When electric waves impinge upon the coherer, its resistance falls from a nearly infinite value to something between 500 and 100 ohms, which allows the cell, g, to energize the electro-magnet of the relay, n, and close a circuit containing a larger battery, r, together with a Morse recorder, h, and a trembling electric bell, p, to act as decoherer. The hammer, o, of the bell is so adjusted as to tap the coherer tube and shake the filings in it. If at the moment in which these actions took place the electric waves in the resonator had died away, this tap would restore the coherer to its normal condition of practically infinite resistance, and a dot only would be recorded on the tape of the Morse machine. If, however, the key of the transmitter were kept depressed, then waves would succeed each other at very short intervals, so that the acquired conductivity of the coherer would only be momentarily destroyed by the tap of the bell hammer, and immediately re-established by the electric waves.

Small choking coils, k^1 k^1—that is to say, coils wound so as to have self-induction or electric inertia—are introduced between the coherer and the relay, their effect being to compel the greater part of the oscillatory current induced in the circuit by the electric waves to traverse the coherer, instead of wasting the greater portion of its energy in the alternative path afforded by the relay.

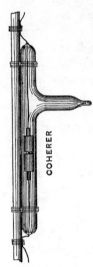

434. WIRELESS TELEGRAPHY. Marconi coherer. The most important part of the receiver is the coherer, which consists of a small glass tube about two and a half millimeters in internal diameter and some four centimeters in length. Two silver pole pieces are lightly fitted into this tube, separated by a gap of about a millimeter, containing a mixture of 96 parts of nickel and 4 parts of silver, not too finely granulated, and worked up with the merest trace of mercury. This powder must not be packed too tight, or the action will be irregular and oversensitive to slight outside disturbances, while if too loose it will not be sufficiently sensitive. It is found that the best adjustment is obtained when the coherer works well under the actions of the sparks from a small electric trembler placed at a distance of about a meter. The tube is then exhausted on a mercury pump until the pressure falls to about a millimeter, when the tubulure left for exhausting it is sealed off. The tubes are tested over a distance of 18 miles before being put into use.

435. WIRELESS TELEGRAPHY. Marconi transmitter with parabolic reflector. When it is desired to send a beam of rays in some

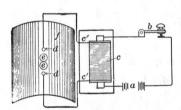

definite direction, the transmitter used by Marconi is one devised by Professor Righi, of Bologna. The two large spheres, *e, e,* are 11 centimeters in diameter, and are separated by a space of a millimeter. In order to concentrate the beam of rays in the required direction the oscillator is placed in the focal line of a parabolic cylindrical reflector. *f,* parabolic reflector; *c, c', c',* induction coil; *a,* battery; *b,* key.

436. AUTOMATIC TROLLEY-WHEEL GUARD. The trolley wheel is linked to a fork and to a counter-weight on a lever by a

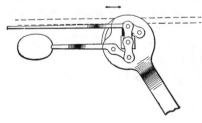

sliding journal box. At the moment the trolley wheel leaves the conducting wire, the weight on the lever lifts the wheel and fork, which again fall on contact of the wheel with the conductor.

437. WIRELESS TELEGRAPHY. Long-distance Marconi transmitter, when it is not required to concentrate the waves in one direction.

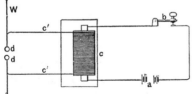

The small spheres, d, d, are connected by the wires, c', c', with the secondary terminals of an induction coil, c, and one of them is also connected with the vertical wire, W, while the other is earth-connected. When the Morse key, b, is depressed, the coil is energized by the battery, a, and therefore, as long as the key is operated, a stream of sparks is maintained between the spheres, d, d.

438. ELECTRIC LIGHTING SYSTEM. A is the alternator, generating, say, 1,000 volts. The lamps used for ordinary illumination,

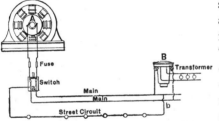

such as for residences, etc., are connected through transformers, as shown at B. The street-lighting circuit consists of a number of incandescent lamps l, all connected in series and cut in across the mains at a, b; the point b may be at the station or on the line, whichever is the more convenient. In order that such a series system may work successfully, the current in the circuit must be kept at a certain value, for which the lamps are designed. It is also evident that there must be a sufficient number of lamps connected in series to take up the voltage of the dynamo. For example, if each lamp required 20 volts, there would necessarily be 50 lamps connected in the circuit, unless some outside device, such as a resistance or choking coil, was used to take up the extra voltage.

439. ELECTRICALLY HEATED CHAFING DISH. The cylindrical box under the dish is fitted with a resistance coil of iron wire insulated with asbestos packing.

440. VIBRATING ELECTRIC BELL. A spring, R, is attached at T to a fixed metallic rod, and presses against the rod T¹. The current

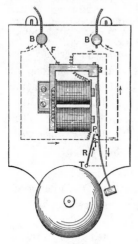

enters through the terminal, B, traverses the bobbins, passes through T, through the spring, through T¹, and makes its exit through the other terminal. The armature is attracted, and the point, P, fixed thereto draws back the spring from the rod, T¹, and interrupts the current; but at the moment at which the point touches the spring, and before the latter has been detached from the rod, T¹, the electro-magnet becomes included in a short circuit, and the line current, instead of passing through the bobbins, passes through the wire, T, the armature and the rod T¹. The vibration of the armature breaks the contact at T¹.

441. PRINTING TELEGRAPH. The type wheel, *j*, driven by clockwork mechanism from the spring barrel, is placed on a shaft con-

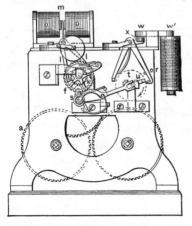

centric with the ratchet, *k*, which is controlled by pallets of the escapement, *l l*, attached to the permanently magnetized armature, *m*, vibrated by alternating currents through the electro-magnets, *o o*. The electro-magnet, *r*, controlling the printing escapement, is in the same circuit, its core having an extension, *w*, and, being surrounded by a non-magnetic material, it is not operated by the rapidly changing currents passing through *o o*, which work the type-wheel escapement; but when the key connected with any particular letter is struck, the circuit is closed, sufficient magnetism accumulates in the core, *w'*, to attract the armature, *x*, releasing the arm, *u*, carrying the paper-roller, *s*, and allowing a crank pin on the shaft of the wheel, *t*, turned by a train of mechanism from the spring barrel, *a*, to depress that arm of the lever and throw the feed roller up against the type.

442. ELECTRIC FIRE-ALARM SYSTEM, Jersey City, N. J. Signal post and call box. The bells are rung by mechanism actuated by a descending weight of 3,000 lb. When on closed circuit, the armature is held by the magnet and the motion is arrested. When the circuit is opened, the armature falls back from the magnet. This releases the detent, and the ratchet wheel holding the weight begins to revolve. Referring to the cut, it will be seen that there are two pawls which engage with the teeth of this ratchet wheel. Each pawl is held to its position in engagement with the teeth of the wheel, or is released therefrom by the action of a pin projecting at right angles from the pawl, and projecting through a slot of peculiar outline. This is shown in the cut directly below the drawing of the ratchet wheel and weight. This slot and pin mechanism is so arranged that only one of the pawls at a time engages with the teeth. When on closed circuit, the upper pawl only is in engagement. When the detent is released, the upper pawl is first acted on by the revolving wheel. This action draws the hammer back from the bell. As the pin rides through the slot, the pawl escapes from the teeth, the other one engages, and the hammer is driven against the bell. The bell-ringing lever rises and is again caught by the detent just as the pawls change places, and the motion is arrested with the upper pawl engaged until the next break in the current occurs.

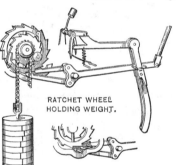

RATCHET WHEEL HOLDING WEIGHT.

443. Striking mechanism.

444. Electric connection and hammer.

445. ELECTRIC TREE-FELLING MACHINE. The two-wheeled vehicle is anchored to a tree ; the motor on a platform drives

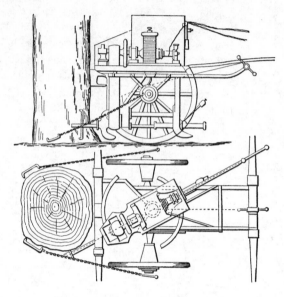

by belt a routing tool below, both swinging on a common center. A handle extending to the rear serves to guide the cutting tool, and a ratchet and rack feeds the cut. German design.

446. Plan, showing grappling chains, routing tool, frame and handle.

447. ELECTRIC TRUMPET. The apparatus consists of a brass tube $2\frac{1}{4}$ in. in length and $1\frac{1}{2}$ in. in diameter, in the interior of which

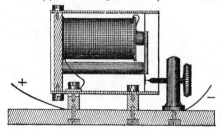

is fixed a small electro-magnet. An armature is placed opposite the poles of this latter, and a regulating screw terminating in a platinum point serves as an automatic interrupter. It takes but two Leclanche elements of the usual electric-bell variety to cause it to produce an agreeable musical sound, of which the pitch and intensity may be varied by regulating the screw or tightening up the vibrating plate in its setting.

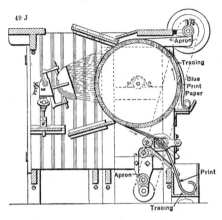

448. ELECTRIC BLUE PRINT MACHINE. The making of blue prints by electric light has been carried to the point where plate glass disappears from the apparatus used. The machine consists of a large wooden cylinder, which is made to revolve slowly in front of the lamp, and any good photo-engraver's lamp will do. A transparent traveling apron moves with the drum. The apron is reeled up on a small drum at the bottom of the machine, and this and the upper roller upon which it is wound keep it in tension sufficient to always hold the tracing and printing paper close together and against the large drum. The tracing and sensitized paper are fed in under the moving transparent apron at the top, and both are received in a box placed below the large drum. The driving mechanism can be operated by belt from shop shafting or run by a small electric motor. The whole apparatus may be arranged so as to receive sunlight upon fine days by being mounted upon a truck.

449. DEMAGNETIZING A WATCH. The center of the watch, C, is placed so that the prolongation of the axes of the magnet (shown

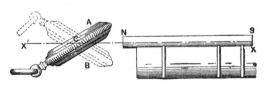

by the dotted line, X X') passes through it. The watch is vibrated around an axis passing through C and at right angles to X X'. By this operation the watch is successively brought into the positions A and B, in all positions around the hour circle.

A few minutes of this movement is sufficient to demagnetize a watch.

450. ELECTRIC CURLING-IRON HEATER. A resistance coil of iron wire insulated by asbestos, inclosed in a brass tube, to receive the curling tongs.

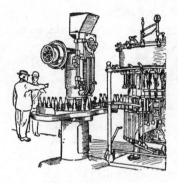

450A. ELECTRIC CAPPING MACHINE. The great increase in the consumption of bottled beverages has resulted in the building of electrically driven machines for doing many of the operations in this field, among them being the machine here shown for attaching the crown caps upon bottles. This device caps 100 bottles a minute and does not require an operator as the bottles are fed directly from the filling machine to the capping device.

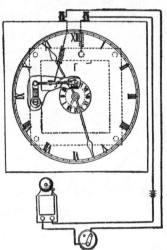

450B. ELECTRIC TIME SIGNAL. This signaling apparatus consists of a signal bell and battery circuit, one terminal of which is connected to a brush bearing against a disk of insulated material, while the other is connected to a segment in the aforesaid disk. The disk is mounted on the cannon of the hour hand, and connected therewith is a small dial on the face of the clock. To set the alarm for any desired hour, the dial is turned so that the brush will come in contact with the metal segment when the hour hand of the clock reaches the desired hour.

SECTION XI.

NAVIGATION, VESSELS, MARINE APPLIANCES, ETC.

187

NAVIGATION, VESSELS, MARINE APPLIANCES, ETC.

451. CURIOUS BOATS. Skin boat of the Gros Ventres Indians, Dakota. An ash or hickory withe frame covered with rawhide. A type of the odd Welsh coracle.

452. CURIOUS BOATS. Pernambuco, Brazil, catamaran, with a caboose and plat-form on a frame above log float.

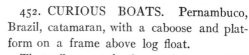

The gaff sets at the foot of the mast, which is stayed by cleats and braces to a cross piece pinned to the logs.

453. CURIOUS BOATS; GREENLAND KAYAK. Has no keel, made of sealskins, and entirely covered except a space to slip the body of the occupant into a seat on the bottom.

454. CURIOUS BOATS. The sheltered duck boat.

The canvas wings drawn back from a short mast make a hiding-place for the gunners.

455. CURIOUS BOATS. A Norwegian fishing smack. A type of the Danish Vikings' war boats. A gaff, but no boom.

Type of a thousand years.

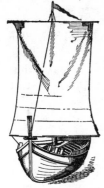

456. CURIOUS BOATS. Reef-sail yacht of Norway. A Vikings' model with square sail.

The Viking types have no deck; the mast is stepped on the keelson and braced from sides and keelson; a jib stay but no jib.

457. CURIOUS BOATS. The Dutch fishing pink. The boats are nearly as broad as they are long. The mast is stepped nearly in the middle of the boat. Two bowsprits with two jibs and an overboard leeboard of rude construction seem to hold these bulky boats to their course.

458. CURIOUS BOATS. The Philippine Island catamaran and anchor. The boats are so narrow that outriggers and floats are required to bear any sail.

A type of the boats of the Pacific islanders in which daring surf feats are performed.

459. CURIOUS BOATS. A Russian canoe rig. Boom and gaff extending forward. Forward part of boat covered. Halyards extend to cockpit for easy management by one person. The boom is elastic and bends before the wind ; a fad of doubtful efficiency.

460. CURIOUS BOATS. The Sandwich Islands catamaran. A narrow boat with outrigged platforms and latticed extension carrying a balance float.

Shrouds from mast to outrigger. Boom and gaff extending forward.

461. CURIOUS BOATS. A Bombay yacht, with Malay rig. A curiously formed bottom. This rig is of the latteen type of the Mediterranean, with short masts inclined forward, yards hung at the center of the wind pressure.

462. C U R I O U S
BOATS.   The turkey-
bone yacht with a swing-
ing bowsprit—more curi-
ous than useful.

463. CURIOUS BOATS.   Non-heeling
sail-boat. A heavy keel frame is pivoted at
stern and bow to which the mast is fixed.

This rig allows the mast and sail to lean
from the wind, while the boat is balanced by
the contrary swing of the iron keel.

464. RACING YACHTS.   Model designs of the British and American yachts contesting for the Victoria cup in the international races
since 1885.   Hull and midship section.

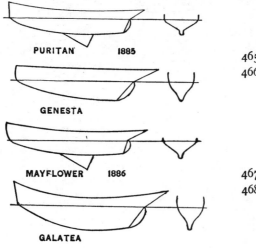

PURITAN   1885

GENESTA

MAYFLOWER   1886

GALATEA

465. Puritan.
466. Genesta.

467. Mayflower.
468. Galatea.

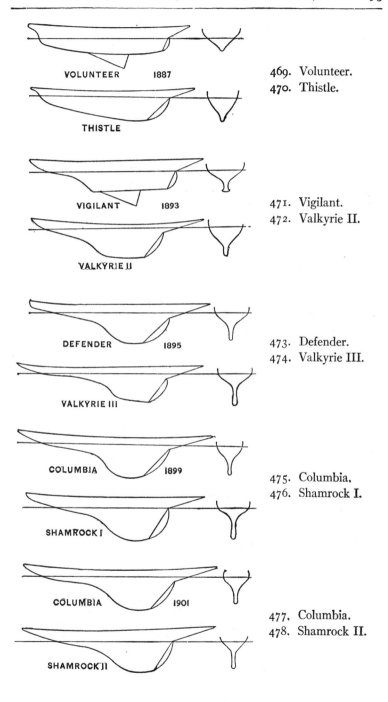

469. Volunteer.
470. Thistle.

471. Vigilant.
472. Valkyrie II.

473. Defender.
474. Valkyrie III.

475. Columbia.
476. Shamrock I.

477. Columbia.
478. Shamrock II.

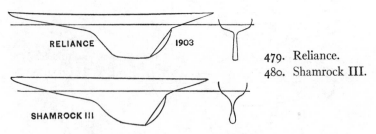

479. Reliance.
480. Shamrock III.

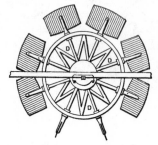

481. ANCIENT FEATHERING PADDLE WHEEL. Wipers on the inner ends of the paddle arms, rubbing against a fixed cam plate, turn the paddles as they enter the water.

482. TYPES OF PROPELLERS. Thornicroft propeller, used on fast boats with fine lines. Blades broad on the hub, narrowing toward the outer end of the blade, face a parabolic recess. Two or three blades. Pitch of blades at point two and one-half times diameter of propeller.

483. TYPES OF PROPELLERS. The Jarrow propeller, with two or three blades curving backward and narrowing from hub to point. A high speed propeller. Face of blades with recess curves and pitch at tips about two and one-half diameters. For fine line boats.

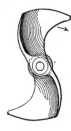

484. TYPES OF PROPELLERS. The Hirsh propeller. The generating line is the segment of an Archimedes spiral, with the leading edge of the blades curved forward, face of the blades curved with increasing pitch.

485. SCREW PROPELLER. Three blade. Reeves type. Curved to conform to an even thrust in all parts of the blades. Good form for launches. Pitch about twice the diameter.

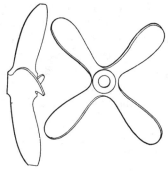

486. SCREW PROPELLER. Four blades. Case type. With outward thrust. Narrow blades for high speed. Pitch two and a half times the diameter. Face of blades curved.

487. Plan of propeller in the plane of rotation.

488. SHEET METAL PROPELLER. Davis type. The blades are made of boiler plate, or of plate steel, of equal thickness through-

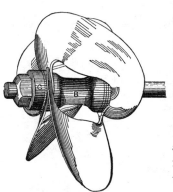

out. They are cut from a flat plate, the holes for the reception of the propeller shaft made, and then either by hammer, rolls, or formers curved to the proper shape. Each blade is precisely alike, so that if one should be broken a duplicate could be readily fitted.

A collar is secured upon the shaft, and the inner legs of the blades bear firmly against it. The sleeve, B, keeps the legs of the blades at the proper distance apart, and the collar, C, and nut secures all in place. To hold the blades in position against the leverage of the water, bolts may be passed through the collars and blades longitudinally with the shaft, or the blades may be held by a feather on the shaft.

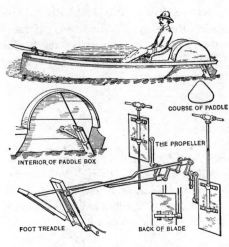

COURSE OF PADDLE

THE PROPELLER

INTERIOR OF PADDLE BOX

FOOT TREADLE   BACK OF BLADE

489. FEATHERING BLADE PROPELLER. A foot-power propeller with paddle blades to hang over the sides at the stern, or may be placed in an extension at the stern, as shown in the cuts. The longest movement of the paddle is when it is immersed, and the paddle being vertical, there is no splash, slip, or loss of propulsive effect arising from the oblique action.

490. The curve traversed by the edge of the paddle.

491. The extension of the paddle box.

492. Crank shaft and foot treadle connections.

493. Blade and crank connection.

494. TWENTY-FIVE-FOOT LAUNCH. Fast type with positive submerged wheel, 5 feet beam, light draught hull with wheel depth of

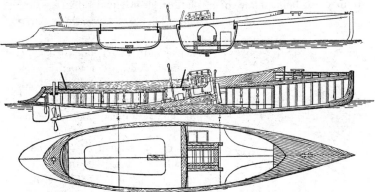

26 inches. Wheel 20 inches diameter, 30 inch pitch, with or without steel wheel guard. Motor, 12 horse power. Displacement with passengers, 2,000 lbs. Elevation and midship sections.

495. Sectional elevation of frame, motor and propeller.

496. Plan with lines of upper works.

497. BICYCLE CATAMARAN. The pedal shaft carries a large worm gear meshed to a small worm gear on the propeller shaft—steering is by the bicycle handle and cross arm below, with wire lanyards to the rudder.

498. BICYCLE GEAR FOR A BOAT. The sprocket wheel shaft with a large bevel gear drives a vertical shaft with two bevel gears

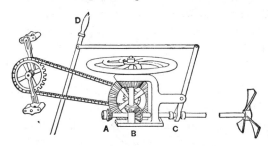

and a fly wheel. The propeller shaft has two bevel gears to mesh into the bevel gear on the vertical shaft alternately, for forward or backward motion of the propeller. Lever C and handle D control the fore and aft motion of the shaft.

499. THE MANIPEDE CATAMARAN. Operated by feet or hands on levers with sprocket wheel and chain to a paddle wheel.

Steered by a rocking seat, or by hand.

500. TYPES OF SHALLOW-DRAUGHT SCREW-PROPELLED BOATS. Yarrow type. Vessels of this class have been constructed

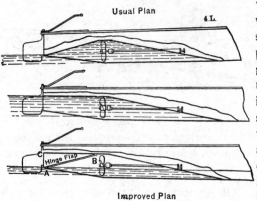

Usual Plan

Improved Plan

with the propeller working in a tunnel, so that though the propeller used be greater in diameter than the draught, yet it always works in solid water, since the water, owing to the air being driven out of the tunnel by the action of the screw, rises and completely covers the latter. Until quite recently the stern portion of this tunnel has been fixed so that when the boat is heavily loaded there is a considerable portion of the back of the tunnel against which the water delivered from the propeller must impinge, and down which it must slide before it can escape. This has caused a "drag," and a consequent loss of speed and efficiency.

In the new type, instead of being fixed, the rear portion of the tunnel is formed of a hinged flap provided with strips of rubber at the side, which, by rubbing against the parallel sides of the tunnel, prevent the ingress of air. With a boat so provided, the rear end of the tunnel can be always so arranged that it only just dips beneath the surface of the water. This is quite sufficient to insure the screw always working in water, but under all circumstances opposes a minimum obstruction to the escape of the water.

501. Section with hinge flap raised for deep draught.

502. Section with hinge flap down for shallow draught.

503. DIRIGIBLE TORPEDO. Sims-Edison type. The front compartment contains a charge of from 250 to 500 lbs. of high ex-

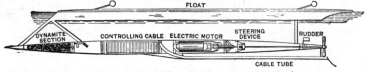

plosive, which can be exploded electrically by reversing the current. In another compartment is a reel upon which is stowed from one to two miles of controlling cable. The cable is made extremely light and flexible, but of sufficient area to convey the 30 horse-power necessary to drive the torpedo at a speed of 22 miles an hour.

The cable, which is connected with a dynamo at the firing station, is led out through a tube running parallel with the axis of the torpedo to a point aft of, and below, the propeller wheel.

504. AUTOMOBILE TORPEDO. Whitehead type. With self-contained motive power for short range action. A 3-cylinder motor

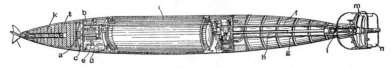

driven by compressed air contained in the cylindrical tank i ; m, two screw propellers driven in opposite directions by reversing bevel gears to keep the torpedo from turning over ; h, shaft ; g, j, steering-apparatus connections from electric steering gear, a, b, c ; k, fuse ; t, explosive charge ; n, rudder.

505. THE HOLLAND SUBMARINE BOAT. A, torpedo tube for rear discharge ; B, dynamo ; C, gasoline engine ; D, air compressor ;

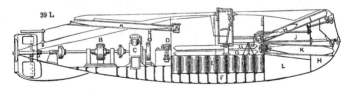

E, storage battery ; F, oil and water tanks ; G, compressed air chambers ; H, gun-cotton shell in the aerial torpedo gun ; J, magazine ; K, Whitehead torpedo and tube ; L, trimming tank, oil and gasoline tank at H under the bow.

In the Holland submarine boat the gasoline engine and the dynamo are directly connected to the propeller shaft, so that when the boat is running on the surface the gasoline engine is used for power, and when submerged the dynamo alone is in use with current from the storage batteries. The air compressor charges the long air tubes, G, G, to a high pressure before going into action, which is discharged in jets when needed for ventilation and cooling the interior of the boat, and also for discharging the aerial and submerged torpedoes.

506. REVERSING CLUTCH for a launch. C is the propeller shaft ; D, the engine shaft ; A, H, shell keyed to the propeller shaft,

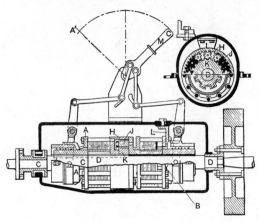

with an inside spur gear to match the ring of pinions J ; L, anchor knee to hold the reverse motion shell B ; K, the inner gear which drives the back motion when the pinion ring is held fast by the position of the lever M. The arrangement of the gear is shown in the cross section.

507. ICE BOAT. Plan and elevation drawn to a scale, as shown in the engraving. Figured measures are given for the most important parts.

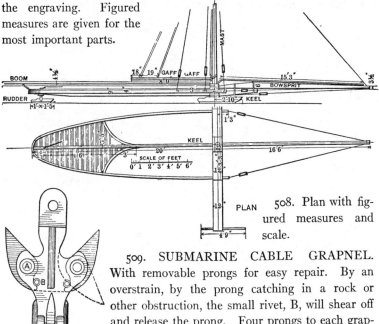

508. Plan with figured measures and scale.

509. SUBMARINE CABLE GRAPNEL. With removable prongs for easy repair. By an overstrain, by the prong catching in a rock or other obstruction, the small rivet, B, will shear off and release the prong. Four prongs to each grapnel. Much used in the repair of defective or broken submarine cables.

510. SUBMARINE CABLE GRAPNEL. Cutting and holding grapnel. The cable on being hooked and lifted forces its way through

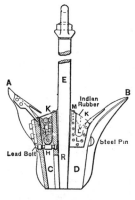

the rubber shield, K, which is provided to keep the mechanism clear from stones, sand, etc., while towing over the bottom, and becomes inclosed in the jaws of the hinged clip. As the strain increases the hinged clip shears through a leaden bolt, H, which supports it, and, moving upon its pivot, is forced down the tapering sides which press the sides of the clip together so as to grasp the cable very tightly. The greater the strain on the grapnel rope, the more the clips are forced down and the tighter the cable is held ; until at last the clips sink so far as to cause the cable to make an acute angle over the knife edge, L, and the cable is cut, one end falling to the bottom while the other is brought to the surface.

511. STEAM SOUNDING MACHINE. Sigsbee type. Its principal parts are the drum, A, on which is wound the wire, the auxiliary

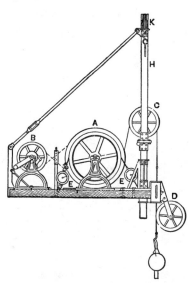

pulley, B, used while heaving in to relieve drum of the strain, the jockey wheel, C, the swivel pulley, D, the accumulator contained in the tube, H, and the brake E.

The drum is made light, in order to have as little inertia and momentum to overcome as possible. Its circumference is one fathom. An indicator is attached to the axle, which registers the number of revolutions. The auxiliary pulley, B, is composed of three pulleys : one for the wire, one for the belt going to the drum, and the other for the belt from the driving engine. The jockey wheel, C, is an ordinary gun metal one with a V-shaped score, and the wire passes over this both in paying out and reeling in. Its circumference is 3 feet, and an odometer being attached to its axle, the amount of wire paid out can thus be obtained. A very important feature in this machine is the accumulator, which is composed of spiral springs contained in two vertical tubes, one of which is shown at H. These springs are connected with the crosshead of the jockey wheel by means of chains passing over the pulley K. The crosshead moves in steel slides, and rises and falls as the weight on the wire varies, indicating on a scale the strain in pounds.

512. THE DRAG STEERING GEAR. Different rigs for constructing and operating a temporary steering gear when the rudder is

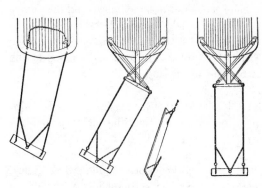

disabled. The float is a strong plank, so fastened by the rope harness as to keep it in a vertical position in the sea.

513. Tackle rig from a projecting spar and cross tree.

514. Rope hitch to the steering plank.

515. Drag gear straight astern.

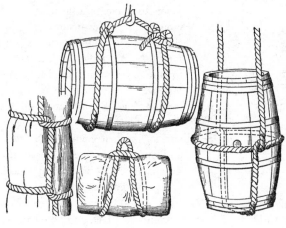

516. ROPE HITCHES. Showing approved methods of hitch for hoisting goods.

517. Hammock hitch.

518. Cask sling and hitch.

519. Bale sling and butt sling on end.

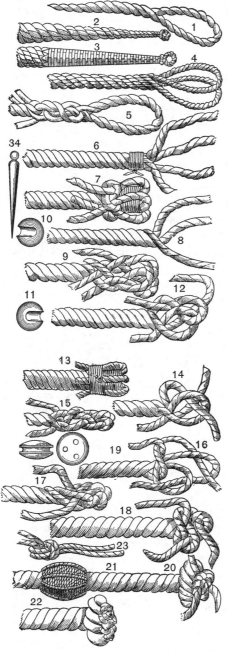

520. KNOTS AND SPLICES:

1. Turn used in making up ropes.

2. End tapered for the purpose of passing it readily through a loop. To make this, unlay the rope for the necessary length, reducing a rope diminishing in diameter toward the end, which is finished by interlacing the ends without cutting them, as it would weaken the work; it is lastly "whipped" with small twine.

3. Tapered end covered with interlaced cordage for the purpose of making it stronger. This is done with very small twine, attached at one end to the small eye, and at the other to the strands of the rope, thus making a strong "webbing" around the end.

4. Double turn used for making rope.

5. Eye splice. The strands of the cable are brought back over themselves, and interlaced with their original turns as in a splice.

6. Tie for the end of a four-strand rope.

7. The same completed, the strands are tied together, forming loops laying one over the other.

8. Commencement for making the end by interlacing the strands.

9. Interlacing complete, but not fastened.

10 and 11. Shell in two views, showing the disposition at the throat.

12. Interlacing in two directions.

13. Mode of finishing the end by several turns of the twine continued over the cable.

14. Interlacing commenced in one direction.

15. Interlacing finished, the ends being worked under the strands, as in a splice.

16. Pigtail commenced.

17. Interlacing fastened.

18. Pigtail with the strands taut.

19. Dead-eye, shown in two views.

20. Pigtail finished. We pass the ends of the strands, one under the other, in the same way as if we were making a pudding splice, thus bringing it in a line with the rope, to which it is seized fast, and the ends cut off.

21. Scull pigtail; instead of holding the ends by a tie, we interlace them again, as in No. 16, the one under the other.

22. Pigtail or "lark's nest."

23. Two-strand knot.

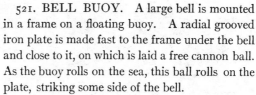

521. BELL BUOY. A large bell is mounted in a frame on a floating buoy. A radial grooved iron plate is made fast to the frame under the bell and close to it, on which is laid a free cannon ball. As the buoy rolls on the sea, this ball rolls on the plate, striking some side of the bell.

In this design a very small roll of the sea makes a constant ringing of the bell.

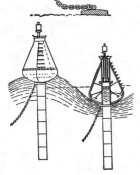

522. THE WHISTLING BUOY. The hanging tube below the float is open at the bottom. In the vertical motion of the float and tube by the waves, the water in the tube reacts as a piston, drawing in air at the top of the buoy and compressing it to blow the whistle.

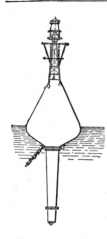

523. LIGHTING BUOY. Compressed gas is charged into the body of the buoy at from 100 to 200 pounds per square inch. A regulator delivers the gas to the burner at a uniform pressure. A single charge will burn for several days.

The inverted cone under the lamp protects it from the splash of the waves. Good for harbors and channels.

524. FOG WHISTLE. A signal of warning operated by wave motion. A sounder on the principle of the steam whistle is exposed to a

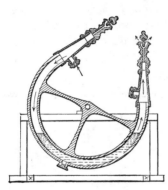

blast of air, according to the facilities of operation. Usually, motion derived from the waves, the tide, the wind, or clockwork, makes it automatic. In the example, the semicircular tubular vessel is mounted upon a rock shaft, and has at each extremity an ordinary whistle and a valve opening inward. When the vessel is partially filled with water and rocked to and fro, the air is forced through the whistle and sounds an alarm.

525. FISH WAY. A device to enable fish to ascend falls or dams. It may consist of a series of stepped basins over which the water de-

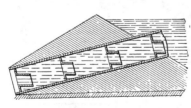

scends, turning a fall into a cascade, and sometimes known as a fish ladder; or it may consist of a chute with a sinuous track for diminishing the velocity and assisting the passage of the fish to the level above the dam. In the example it is an inclined chute having a series of chambers containing comparatively still water, the current being confined to a relatively smaller space.

526. FLOATING BREAKWATER. Morris type. A A are air-tight cylinders ; B B the strutting ; C C the cables, and D D the weights

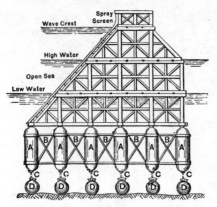

at the sea bed. From the motionless foundation thus formed, the framing rises through the section of tidal and superficial action. The sloping screen formed by the timbers presents meshes to the waves, by which their force is arrested and their effect destroyed. The first idea of floating breakwaters was probably taken from an observation of the effect produced upon waves by the presence of some natural obstacle in the sea, such as reeds and sea weed. The gulf weed is a well-known instance. It has been found that, although its depth does not exceed a couple of feet, yet, even in strong gales, there is perfectly calm water to leeward of it. The illustration represents a form of construction for ocean shields, breakwaters, piers, harbors, gun-banks, lighthouses, and other marine objects.

527. NETS AND SEINES. How they are made. A and *b*, two styles of netting needles, *e*, mesh peg, *ƒ*, flat mesh peg.

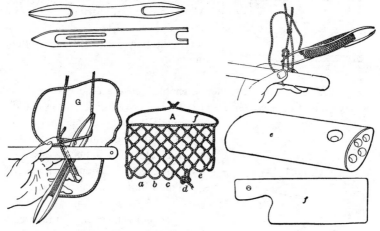

A, section of net, showing last loops at *a, b, c, e*, and the formation of the knot at *d*, with ·the mesh peg left out.

G and Z, unlettered, show the formation of the knot with peg and needle.

528. Closed point needle, American type.

529. Making a loop with open end needle and peg.

530. Oval mesh peg.

531. Flat mesh peg.

532. Section of net, with knot at *d.*

533. Making a loop, second stage.

533A. A HAND-OPERATED TROLLEY FERRY. A unique overhead trolley is shown above. The little four-seated car, suspended from a single wire, travels across the river a few feet above the water at a speed approximating 20 miles an hour, the power being provided by the ferryman turning a wheel over which the hauling cable is wound. The wheel, it will be noticed, has a grooved rim.

533B. BOAT PROPELLER which may be easily and quickly applied to or removed from the transom of a boat. This adapts the mechanism particularly to boats which are to be launched frequently, and the principal advantage

lies in the fact that when the boat is hauled up on the shore or in the davits of a vessel the propelling device may be removed from the transom and stowed safely away.

533C. WATER BICYCLE. The bicycle is closely imitated in this swimming apparatus. Both foot pedals and hand cranks are used to drive the propeller and speed is necessary to maintain its buoyancy, for its only supports are short planes on each side of the body and a cut-water in front. The rudder is placed under the front end of the frame and is controlled by the rider's chin.

533D. SNOW AUTOMOBILE. The car body is of automobile type, but on each side, in place of automobile wheels, is an endless

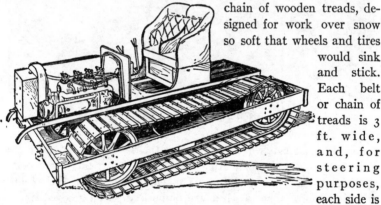

chain of wooden treads, designed for work over snow so soft that wheels and tires would sink and stick. Each belt or chain of treads is 3 ft. wide, and, for steering purposes, each side is under separate control, so that one belt can be stopped while the other carries the machine around a corner. In addition to this there is a similar tread projecting out from the rear of the car, designed to help in steering it.

SECTION XII.

ROAD AND VEHICLE DEVICES, ETC.

Section XII.

ROAD AND VEHICLE DEVICES, ETC.

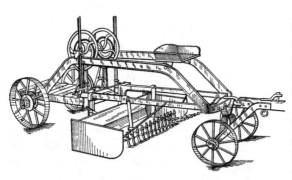

534. ROAD GRADING WAGON. With shifting tongue device on the frame to allow of close scraping on each side of the road. An elevating gear for the rake and scraper.

535. TRACTION WHEEL. In this wheel the projections of the rim are yieldingly held, or they may be withdrawn entirely from the surface or held locked in outermost position.

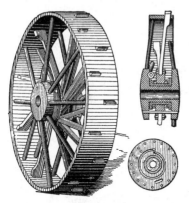

A sleeve fitted loosely on the hub between the flanges carries a loosely rotating wheel on which are pivoted the inner ends of slidable arms, whose outer ends are beveled and pass through openings in the rim. To fasten the sliding arms in either an inner or outer position a pin is passed transversely through apertures in the hub flanges and through one of several apertures in the wheel on which the sliding arms are pivoted, the wheel being turned to the proper position before inserting the pin, while the passing of a pin through the hub flanges and an elongated aperture in the wheel restricts the latter to a limited turning in either direction.

536. Section of wheel, hub flange and slidable arm.

537. Rotating hub with pin slot.

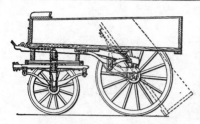

538. DUMPING WAGON. The loaded box is just overbalanced to tip backward. A dog catch on the driver's footboard is let go for self-tipping of the load.

539. DIFFERENTIAL SPEED GEAR for bicycles. Eite & Todd type. A is the crank-axle gear wheel, C the chain wheel, working

through supplementary bracket. On this bracket is a sleeve, B, which carries free running cogs, B_1 and B_2, both running on ball bearings. The chain-wheel axle carries the fixed pinions C_1 and C_2, of different diameter, on one shaft, which are always in gear with both of the pinions, B_1 and B_2. By the action of the lever D, B_1 and C_1, or B_2 and C_2 are thrown into gear with A, thus giving a gear which can be varied in a great range of ratio.

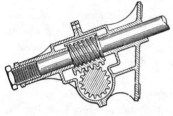

540. AUTOMOBILE STEERING GEAR. The steering shaft has a double thread screw and nut with rack attached, which turns a pinion to operate a shaft and arm connected to the wheel gear. French.

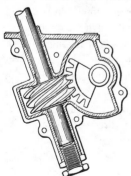

541. AUTOMOBILE STEERING GEAR. A steering shaft with a double thread screw acting on a sector gear, the shaft and arm of which operates the wheel gear. French.

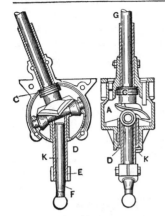

542. AUTOMOBILE STEERING GEAR. A curved and eccentrically mounted cam plate on the handle shaft revolves against roller arms of the hollow shaft K, moving it forward or backward in the socket and sheath D, E. The socket-head spindle, F, accommodates difference in length by sliding in the sheath K.

543. Cross section. French.

544. RATCHET BRAKE LEVER for automobiles. Miller type. By a simple motion of the foot the pawl locks or unlocks the brake lever, so that the brake is on and locked when leaving the automobile alone. Saves much trouble in tying up horseless vehicles.

545. AUTOMOBILE CHANGE SPEED GEAR. Petteler type. A, the driving shaft with fixed gears; B, collar on spear-shaped blade

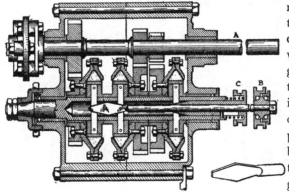

rod for operating the plungers for clutching the forward motion gears; C, collar to a sliding conical sleeve that operates the plungers for the back motion through an idler gear.

546. AUTOMOBILE CHANGE SPEED GEAR. Dorris type.
To the upper shaft are fastened three gears corresponding to the three

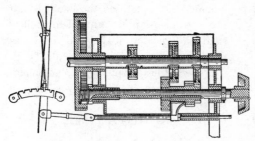

pinions, and in addition an internal gear outside the casing and of comparatively large diameter. A pinion is mounted upon the lower shaft, at the end thereof, adapted to mesh with the internal gear, but is normally held out of mesh by means of a coiled spring at the end of the shaft. The pinion is mounted upon a long sleeve surrounding the shaft and extending through the bearing into the casing. The set of three shifting pinions is shown in the position of slow forward speed. By moving them to the left the second and third speeds are engaged in succession, and after the gears of the third speed are out of mesh, if the motion is still continued, the sliding pinions will abut against the sleeve of the reverse pinion, and shift the pinion into mesh with the internal gear against the pressure of the spring.

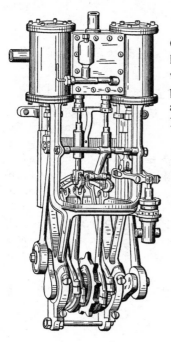

547. AUTOMOBILE STEAM ENGINE. A two-cylinder engine of the locomotive type with link motion and D valves; cylinders $2\frac{1}{2}$ x 4 inches. Boiler pump operated by a lever and link from a crosshead of one of the cylinders. Extreme cut-off o to $\frac{5}{8}$.

The sprocket wheel on the shaft between the eccentrics connects by chain directly with the compensating gear on the rear axle. The prevailing type of engine for all steam automobiles.

DERBY

548. TYPES OF MOTOR BI-CYCLES. The Derby. A chain from the motor drives a friction wheel which is pressed on the tire by a bell-crank lever. This arrange-ment allows of instantaneous motor disconnection.

BROWN

549. TYPES OF MOTOR BICY-CLES. The Brown. Much after the style of the Derby, but driven by a belt from the motor pulley to a pulley attached to the rear wheel.

MINERVA

550. TYPES OF MOTOR BI-CYCLES. The Minerva. The motor hangs beneath the lower reach and drives by belt over a pulley on rear wheel. Has a surface carbureter and tank inclosed in the front frame.

551. TYPES OF MOTOR BICYCLES. The Singer. The motor and all its appurtenances, including fuel tank, are within the

SINGER

rear wheel, which, with the exception of the controlling rods and levers, is inde-pendent of the rest of the bicycle. The motor is hung on a fixed shaft with its crank shaft below the axial center of the wheel, and with a pinion meshing in an internal gear on the wheel. Ignition is by a small magneto.

552. TYPES OF MOTOR BICYCLES. The Humber. The motor is built into the lower reach of the frame in a novel way, com-

HUMBER

prising four tubes as an inclosure. The motor drives a sprocket on the pedal crank shaft by chain, and by another chain to the rear wheel sprocket. A friction disk on the crank shaft pre-vents jerking of the chains under undue strain.

F. N.

553. TYPES OF MOTOR BI-CYCLES. F. N. The motor is clamped in a vertical position in the front frame, with a belt drive to a pulley fastened to the spokes. Motor appurtenances inclosed in a case fitting the upper part of the frame.

554. TYPES OF MOTOR BICYCLES. The Werner. The motor in a vertical position is built into the lower part of the frame and forms part of the frame. The drive is direct by belt from motor pulley to a large pulley fastened to the rim.

WERNER

555. TYPES OF MOTOR BICYCLES. Royal Enfield. The motor is secured to the steering head by bracket clamps. The motor drives direct by a long crossed belt to the rear wheel pulley. The front wheel is provided with a band hub brake, and also one on the rear wheel hub.

ROYAL INFIELD

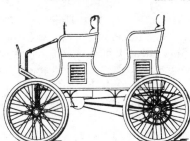

LADY'S IVEL

556. TYPES OF MOTOR BICY-CLES. Ladies' Ivel. The motor is placed beneath the lower front frame and drives by belt to a pulley. Carbureter, igniter, and fuel at the back of the seat post. A skirt shield covers the motor and belt.

557. STEAM SURREY. The boiler is placed under the rear seat and the engine under the front seat, from which the driving by chain is extended to a sprocket on the rear-axle compensating gear. The boiler and engine are illustrated on other pages.

558. STEAM FREIGHT WAGON. Adams Express type. An oil fuel burner under a vertical tube boiler.

Two-cylinder engine directly connected to a two-speed change-gear shaft and to a compensating shaft gear, which in turn is geared inside of the rear wheels.

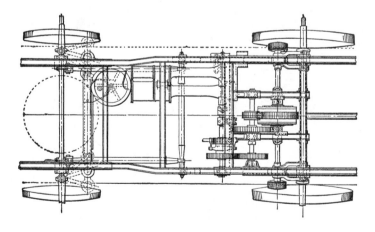

559. Plan of steam freight wagon running gear, with change gear connections.

560. STEAM DRAY. Type of the Leyland dray, much in use in England. Uses a kerosene burner under a vertical tube boiler, with double-reducing chain-gear system. Compensating gear on the reducing shaft.

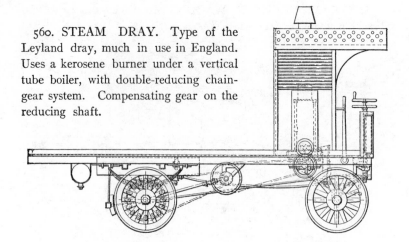

561. INTERCHANGEABLE AUTOMOBILE. A new feature in the combination of a pleasure carriage and a delivery wagon. The passenger entrance is in front. The seat and trim can be readily removed and a hood substituted and the space used for freight.

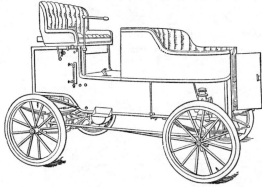

SECTION XIII.

RAILWAY DEVICES AND APPLIANCES, ETC.

Section XIII.

RAILWAY DEVICES AND APPLI-ANCES, ETC.

562. BLOCK AND INTERLOCKING SIGNALS. Electro-pneumatic system. The right-hand figure shows the detail of the air

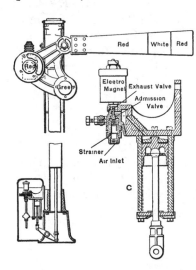

piston and electric air valve. The signal being at the entrance of a block section, which is, say, three-quarters of a mile long, the battery for the current is at the outgoing end; and when the rails of the track, throughout the section (and also the rails of side tracks and crossovers, so far as they foul the main track) are clear—not occupied by wheels at any point—the circuit of the battery is through the right-hand rail of the track to the electro-magnet at the signal, thence to the left-hand rail and by that back to the battery. This circuit being closed, the electro-magnet at the signal is energized and holds the signal, through the medium of a stronger electro-magnet, worked by a local battery, in the *all-clear* or go-ahead position. The entrance of a train short-circuits the current through the wheels and axles, de-energizing the electro-magnet (relay); and the signal, by force of gravity, assumes the stop position, thus warning the next following train not to enter the section. The signal remains at "Stop" until every pair of wheels has passed out of the section.

563. Section showing electro-magnetic valve and pneumatic piston for operating the signal arm.

564. Lever arm connection between air piston and signal arm rod.

565. RAILWAY SIGNALS. The upper cut represents the "home" and "advance" semaphore, and when the blade is placed horizontal

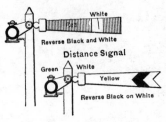

indicates "Danger," or "Stop," and when dropped to the vertical indicates "Clear! Go ahead!" At night the "red light" indicates Danger! the white light Go ahead! The distance signal is placed about 1,800 feet from the home signal. The blade is yellow with a black band, as shown in the lower cut. Its horizontal position by day or a green light by night, indicates "caution."

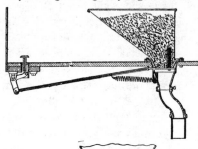

566. TROLLEY-CAR SANDER. A sand box with gate and stirring pin on the gate is operated through the connecting rod by a push button and bell crank.

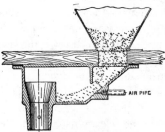

567. LOCOMOTIVE SANDER. A sand box and chute with a nozzle by which compressed air from the air-brake reservoir blows the sand into the discharge pipe.

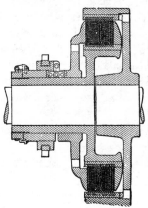

568. MULTIPLE PLATE FRICTION CLUTCH. Pattern of the main driving shaft clutch, Brooklyn Bridge. Every other ring plate is keyed to the inner sleeve and flange; the alternate rings are keyed to driven shaft-flanged hub. A toggle, operated by the collar and a yoke lever, presses the ring plates together for the friction drive.

569. TYPES OF TROLLEY-CAR TRUCKS. Showing different designs of frames and fenders.

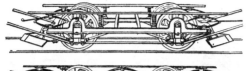

570. Steel cross-bar frame. Leaf springs under car.

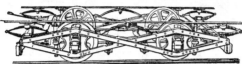

571. Cast - steel box case riveted to wrought iron frame.

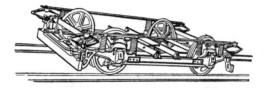

572. Frame supported on spring boxes. Vertical fenders.

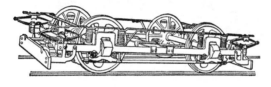

573. Shovel fenders, on spring box frame.

574. Helical spring boxes with leaf springs under car body.

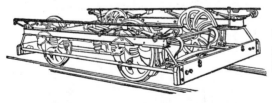

575. Cast - steel box frame bolted to straight iron frame.

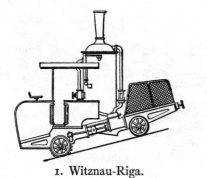

1. Witznau-Riga.

576. TYPES OF RACK-RAILWAY LOCOMO-TIVES for mountain railways. The drive is from the crank, rod, and shaft, with a pinion meshing with a gear wheel on the rack-wheel shaft. Highest grade 1 to 10. Witznau-Riga Railway.

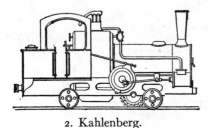

2. Kahlenberg.

577. Locomotive of the Kahlenberg Railway.

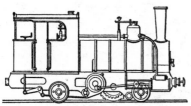

3. Schwabenberg.

578. Locomotive of the Schwabenberg Railway.

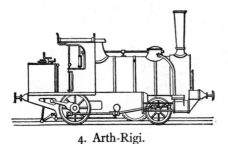

4. Arth-Rigi.

579. Arth-Rigi Locomotive.

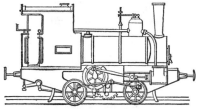

580. Ostermundigen Locomotive.

5. Ostermundigen.

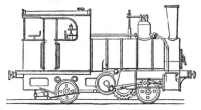

581, 582. Wasseralfingen Railway.

6. Wasseralfingen.

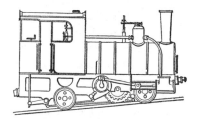

7. Wasseralfingen.

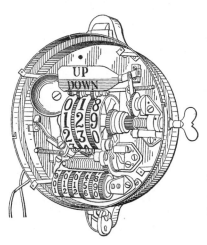

583. FARE-RECORDING REGISTER. Complete with "total" index, trip sign, and bell. Face removed to show the mechanism.

The key at the right returns the trip index to its normal zero, and also sets the "up" and "down" index to its slot in the face. The total index is a continuous register and cannot be tampered with.

584. CABLE GRIP of the Brooklyn Bridge. A plan view from beneath, a section through the sheaves, and a section through the center showing the solid or fixed grip.

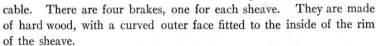

In the grip there are four sheaves placed in pairs, so that the cable is gripped between each pair. Each sheave has a heavy grooved rim with a cylindrical inner surface against which the brake presses. The rim is in two parts bolted together, and holds in a dovetail groove a packing of leather and India-rubber belting in alternating pieces placed radially. The packing projects well out of the rim, and is grooved to receive the cable. There are four brakes, one for each sheave. They are made of hard wood, with a curved outer face fitted to the inside of the rim of the sheave.

585. Cross section of brake frame and cable sheaves holding the cable.

586. Lever links and grip blocks.

587. RAILWAY TRACK BRAKE. By the double toggle joints and lever connection, the whole weight of the truck and end of car is brought to bear on the brake slippers, the lever fulcrum being fixed to the truck frame.

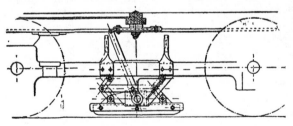

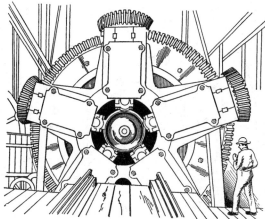

588. ROLLING AND COMPRESSING STEEL CAR WHEELS. Fowler type. Five small wheel tread rolls spaced around the hot car wheel are revolved and pressed to the wheel rim, reducing its diameter a half inch. The inner form of the wheel is kept true during the rolling by molds clamped to each side. The small section shows the clamped wheel. By the rolling process the tread of the wheel is condensed and given the same quality as in steel tires.

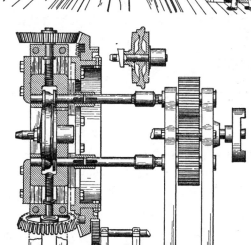

589. Vertical section of frame with wheel between the rollers.

590. REVERSING CAR SEAT. A shifting back seat actuating a foot rest when shifted, to move into proper position to carry it out of the way of the occupant of the seat and leave a baggage space under the seat, while at the same time, by the same movement, the foot rest is properly placed for the occupant of the rear seat. Also to tilt the cushion to the proper level for each way the seat is turned.

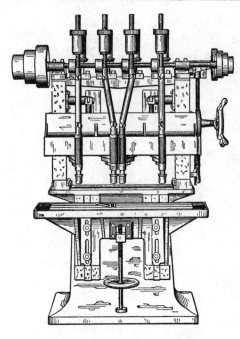

591. FOUR-SPINDLE RAIL DRILL, arranged to drill four ¾-inch holes at once, either in line or staggered.

The distances of the drills are compensated by double universal joint rods.

The drill spindles run in sleeves adjustable on a cross bar, which slides by a hand wheel gear for feeding the drills.

592. CRANK-PIN TURNING MACHINE. The rig comprises essentially two tool carriages turning around the crank shaft, which is

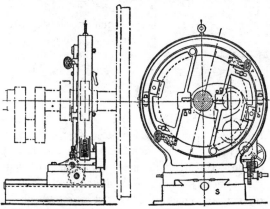

fixed, and it is independent of any special device that may be used for centering the shaft.

The two tool carriages are shaped as circular segments and are diametrically opposite each other, pivoted at one end on a toothed c r o w n which is made in two parts. This crown turns in a circular frame, also made of two pieces, and is driven by a pinion connected with a pulley belted to the shafting of the shop. The circular frame is mounted on a sliding carriage which may be moved with a screw, automatically or by hand, on the saddle S, adapted in size and shape to the lathe bed.

The position of each of the tool carriages may be regulated by turning them on their pivots, bringing them nearer or farther from the axis of the frame which coincides with that of the journal to be trued.

593. Cross section of crank pin and tools set in the tool carriage sectors.

594. EXTENSION CAR STEP. The extension step is carried on a forked arm which slides in guides under the lowest

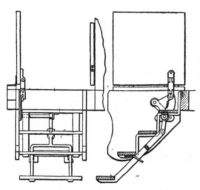

fixed step. The upper end of the arm is connected to a crank arm fixed on a shaft carried in brackets under the top step. On the inside end of this shaft is a toothed wheel which engages with a similarly toothed sector fastened to the face of the step hanger. This sector has an arm on the upper end of the arc to which a link is attached, and the link is in turn fastened to the un-

der side of the vestibule trap door. When the vestibule trap door is closed the crank arm on the shaft is brought to its highest position and the forked arm with the extension step is drawn up close under the fixed step. On raising the trap door preparatory to opening the vestibule doors, the shaft is revolved and the forked arm pushed out, carrying the extension step with it.

595. Side view, with step extended.
596. Side view with step closed.

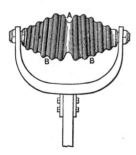

597. TROLLEY REPLACER. The double spiral grooved cone carries a central groove A for the wire, and on each side a helical groove, B, B, which quickly carries the wire to the central groove when displaced. Thus the conductor does not require any special skill in replacing a displaced wheel, for if the wheel catches the wire in any part it is automatically carried to the center groove A.

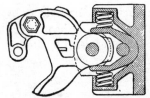

598. CAR COUPLER. Washburn type. Has the side movement of the draw bar, and also a movement of the head of the coupler controlled by the side thrust of the helical springs for centering the coupler head.

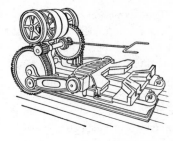

599. BULLDOZER PRESS. For quickly bending straps and braces of iron or steel for car and other constructive work. In this way a large number of forming blocks are used of different designs to fit the slides of the machine.

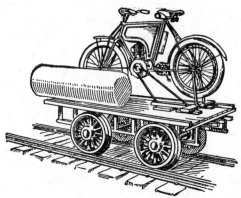

599A. HAND - CAR DRIVEN BY MOTORCYCLE. The 3½ hp. motorcycle is mounted on a 300-lb. railway hand-car, and the drive from the machine is by V-belt to a pulley on a countershaft, and thence by chain to the front wheels of the car. When riding alone, one has to carry ballast, otherwise the wheels would not hold to the track, but with two people aboard the arrangement is ideal.

SECTION XIV.

GEARING AND GEAR MOTION, ETC.

231

SECTION XIV.

GEARING AND GEAR MOTION, ETC.

600. NOVEL WORM GEAR. The threads of a spiral worm, instead of gearing into teeth like those of an ordinary worm wheel, actu-

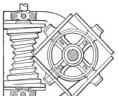

ate a series of rollers turning upon studs, which studs are attached to a wheel whose axis is not parallel to that of the worm, but placed at right angles thereto. When motion is given to the worm then rotation is produced in the roller wheel at a rate proportionable to the pitch of worm and diameter of wheel respectively.

The pitch line of the screw thread forms an arc of a circle whose center coincides with that of the wheel, therefore the thread will always bear fairly against the rollers and maintain rolling contact therewith during the whole of the time each roller is in gear, and by turning the screw in either direction the wheel will rotate.

601. SWASH-PLATE GEARS. The two gears A and B in appearance are two elliptical gears working under the impossible

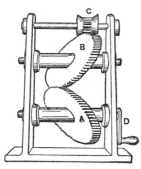

condition of fixed center distances with their major and minor axes coinciding. These gears rotate at the same velocity ratio, and B drives a third spur gear, C, having flanged sides. The gear C is not only rotated but is reciprocated back and forth along its bearing, engaging the sides of its driving gear. It is, of course, obvious that the "elliptical" gears are in reality swash plates or spur gears, formed as a diagonal slice from a spur gear having a length equal to the elliptical section projected on its axis. It will be observed that the teeth are cut parallel with the shafts and all are the same distance from their respective centers, so that the paradox is one of appearance only.

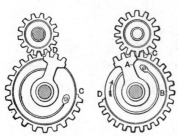

602. STOP-GEAR MOTION. B is the driving gear, with a loose sector, A, held to its forward position by a light spring to catch the teeth of the driven pinion and hold them in position to mesh with the teeth of the driving gear when its stop at D reaches the sector. The stop is during the traverse of open space through which the sector moves.

603. Right hand figure shows commencement of the stop motion, which ends when the stop, D, reaches the sector.

604. VOLUTE TAPPET GEAR. A pinion of the smallest number of teeth, consisting of two spiral teeth so curved that the point of

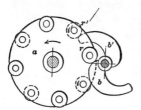

one tooth engages with the friction roller of the next tooth while the preceding roller is engaged with the opposite tooth of the pinion. The alternate roller teeth are on opposite sides of the roller gear, and the pinion teeth are offset to match them.

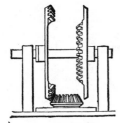

605. GEARED REVERSING MOTION. Broken sections of teeth on a pair of bevel gears alternately reverse the motion of a bevel pinion.

Guide fingers are necessary in this class of gearing for insuring the meshing of the teeth.

606. ELLIPTIC LINKAGE from circular gears. Three equal gears D, G, C, with the linkages A, E, B.

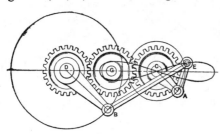

There are many variations of this form of gear and linkage in regard to the forms of curves which may be produced. Arm B, D, is twice the length of arm A, C. Link A, E, is equal to A, C.

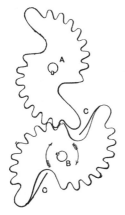

607. INTERRUPTING CAM-GEAR MO-
TION. B, the driver. The motion of A is
from fast to slow or slow to fast, with a momen-
tary stop as the long teeth match at C and C.
The stop motion is governed by the form of the
curves of the long teeth at C and C.

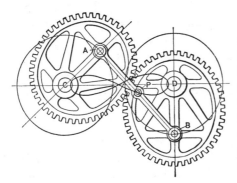

608. ELLIPTIC LINK-
AGE from elliptic gear.
C and D are centers of revolution of the elliptic gears,
and A, B, their opposite focii,
to which the link A, P, B,
is attached. P, the pencil,
which on moving from the
center of the link, will produce a great variety of curves.

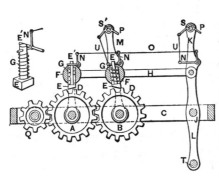

609. CIRCULAR FROM
RECIPROCATING MO-
TION. A lever L, moved by
any reciprocating power, operates the pawls on the meshed
gear A, B, for a continuous
motion of the pinion Q. The
bell-crank levers and connecting rod O are for lifting the
pawls. Suitable for a windmill attachment.

610. Pawl with spring, bell crank and lanyard for lifting the pawl.

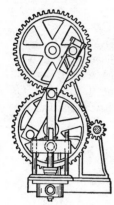

611. CRANK SUBSTITUTE. The gear wheels pinioned to the link, to the center of which the pump rod is pinioned, give a parallel motion to the pump, thus avoiding the lateral thrust of a crank.

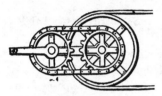

612. SUN AND PLANET MOTION by sprocket wheels and chain. The central sprocket on the pulley shaft is fixed. The belt wheel and its arm carries the second sprocket around with its arm constantly in one direction, which makes its outer end describe a circle eccentric to the driving shaft center. The eccentric circle is not shown in the diagram.

613. INTERMITTENT ROTARY MOTION by a triangular cam on a rotating shaft. The cam works in a yoke forming part of

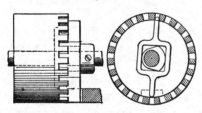

a sliding and vibratory lever dog, the opposite ends of which are adapted to alternately engage with the teeth of the crown-gear wheel, one end of the vibratory lever dog being held by a fulcrum piece or guide so that the cam vibrates the other end.

614. Front view of cam, lever dog and toothed wheel.

615. FRICTION GEAR with cog check to prevent slipping. A smooth running gear.

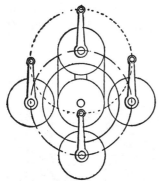

616. PARALLELISM FROM CIRCULAR MOTION. A central pulley which is stationary and belted to a pulley of the same size, but loose on an arm revolving around the stationary pulley, will have an indicator arm on the moving pulley always in the same direction.

An idler pinion on the arm between two equal gears will also produce the same effect on revolving the arm and index wheel around the central gear wheel.

617. CIRCULARLY VIBRATING MOTION. A ring plate pin-

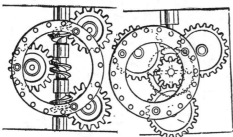

ioned to three gears driven by a central gear, or a right and left screw worm, left hand figure, will swing the ring plate in a circle equal to twice the distance of the wrist pins from the center of the gear wheels.

618. DIFFERENTIAL SPEED GEAR. A speed gearing in which a center pinion driven at a constant rate of speed drives directly and at

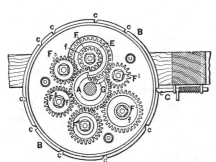

different rates of speed a series of pinions mounted in a surrounding revoluble case or shell, so that by turning the shell one or another of the secondary pinions may be brought into operative relation to the parts to be driven therefrom.

C, a stop for locating each speed pinion.

Each shaft of the pinions, F, F, F, carries below the plate a gear of uniform size, E, which alternately meshes with the driven wheel by the different positions of the shell.

619. EPICYCLIC TRAIN. In which 262,500 revolutions of the left side shaft must be made to produce one revolution in the right side shaft.

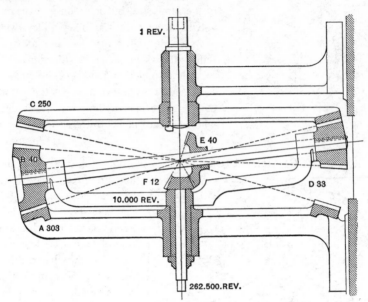

The order of teeth, as marked on the diagram, beginning with the fixed gear A, which has 303 teeth; B, on the cross-arm shaft, 40 teeth; D, at the other end of the cross-arm shaft, 33 teeth; E, also fixed to the cross-arm shaft, 40 teeth; F, on the high-speed shaft, 12 teeth; C, on the slowest wheel shaft, 250 teeth.

620. TRANSMISSION GEAR for automobiles. The three intermediate gears are pinioned to a separate plate from the outside gear, and controlled by a brake strap. There being two compartments and two sets of gear, the brake strap on each compartment controls the speed and the reverse motion.

621. Left hand set of gears.

622. Right hand set of gears.

623. VARIABLE SPEED FRICTION GEAR. The disks on the shaft B are permanently fastened in position by blind screws, and the

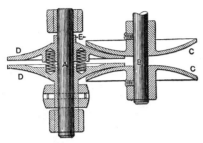

flanges C, C grip two other disks D, D, the latter having springs between them to force them apart and insure a good frictional contact with the disks C, C. The curves are circular arcs with different radii for the two sets of disks, and the design is so worked out that it is but necessary to move the two shafts together or asunder the distance E, about ⅝ inch, to obtain the entire range of the speed variation.

624. VARIABLE SPEED GEAR. In the engraving the shaft, driven direct or through back gears from the pulleys, is shown at *a*. This

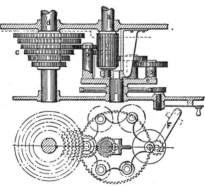

shaft carries at *b* a long pinion. At *c* is a nest of spur gears secured on the shaft *d*, through which the machine mechanism is actuated. Six gears will be noticed in this bank in the revolving gear box. Carried by a rotating frame *e* and meshing with pinion *b* are idler gears, which, as *e* is turned through pinion and gear actuated by means of crank-handle F, mesh one after another with their mating gears in cone *c*, thus giving for each gear so engaged a different rate of speed to shaft *d*.

One turn of crank F suffices to swing one idler out of mesh with its mate and throw the next into action. Hence it is an easy matter for the operator to change the speed even with the machine in motion, as he has only to spring the crank out of a notch which serves to lock it fast in the position indicated and revolve it until the required intermediate is engaged with the gear cone, when the handle is again locked fast by dropping into the notch.

625. Plan of the revolving gear box and handle, F.

626. VARIABLE SPEED GEAR, for automobiles. German type. The gear is of the permanent mesh type, and is adapted to give four

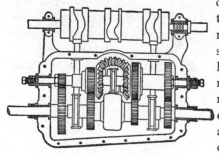

changes of speed and reverse motion. The changes are effected noiselessly and entirely without shock by means of a system of levers actuating friction cones, motion to the levers being transmitted by a series of grooved cams cut from the solid on an auxiliary shaft. The various changes of speed, as well as the reverse motion, are controlled by one lever or wheel which actuates the cam shaft. All the gears are cut from solid steel forgings, and are inclosed in an oil-containing aluminum case. It will be noticed from the drawing that the shaft on which the driven pinions are mounted also carries the differential gear, the shaft being designed to transmit the power by chains to the rear road wheels of the car to which the gear is fitted.

627. DRIVING GEAR FOR A LATHE, change speed gear on two vertical shafts beneath the lathe head. The method of connecting

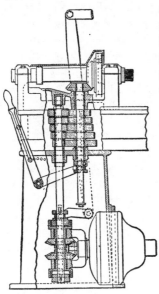

motor and spindle is clearly shown. The armature shaft is fitted with a bevel pinion which drives through bevel reversing gears and a vertical shaft a cone of five gears which mesh with five loose gears on another vertical shaft, the latter being connected by bevel gears with the lathe spindle. By means of a sliding key, operated by a lever at the front of the head, any one of the loose gears may be instantly connected to and made to drive the shaft; thus the speed of the spindle is very readily changed. The spindle is back-geared in the usual manner. The lever for starting, stopping, or reversing the spindle is operated by a rod running the full length of the bed and within convenient reach of the operator.

628. VARIABLE SPEED GEAR. On driving shaft A are secured four spur gears. Shaft B above also has four gears fast upon it. A

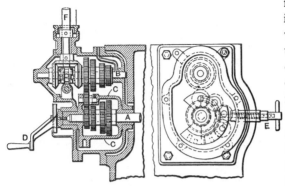

frame or box C, which is mounted in such a way that it may be turned by handle D, carries four intermediate gears which mesh with the driving gear on shaft A. As frame C is turned, any one of the driving gears may be connected—by means of its intermediate—with its mate on shaft B. The index plate on handle D shows how far it should be turned to obtain a certain speed, and when the gears are properly in mesh a spring pin E drops into a hole in the frame and locks it in position. The vertical shaft F can be driven in either direction by means of the bevel gears and clutch on shaft B, the clutch being moved by a lever.

629. Plan of intermediate gears and spring pin.

630. VARIABLE DRIVE MOTION. Two cone pulleys mounted with differential spur gear on the driving shaft with a cross arm and

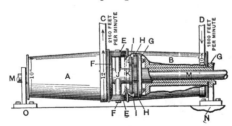

bevel gears, give a variety of speeds between the two cone pulleys. The arm, J, runs loose on the shaft and carries the bevel pinions and a spur gear, K, which operates the differential set of gears H, I.

SECTION XV.

MOTION AND CONTROLLING DEVICES, ETC.

243

Section XV.

MOTION AND CONTROLLING DEVICES, ETC.

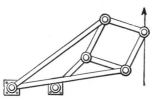

631. PARALLEL MOTION. Peaucellier's seven links. The pivots on the square plates are fixed points. The joint at A makes a straight line. All the short links are equal to the length of the fixed pivots. The other links are three times the length of the short links.

632. PARALLEL MOTION. The three pivots on the square plates are fixed points in a seven-link movement. The links are in

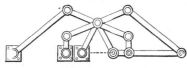

pairs or multiples of pairs. The horizontal bar has a motion on a straight line in the direction of the three fixed points.

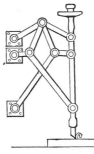

633. PARALLEL MOTION. The three pivots on the square plates are fixed points on an eight-link straight-line movement at right angles to the line of the fixed points. The links are in pairs of equal length.

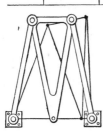

634. THREE-POINT STRAIGHT-LINE LINKAGE. Two links are on fixed pivots and pivoted to the triangular piece at half the length of the fixed pivots distance. The end of the triangular piece carries a tracer on the line of the fixed pivots. At three points in the double curve the tracer crosses a straight line.

245

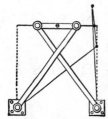

635. THREE - P O I N T STRAIGHT-LINE LINKAGE. The radial bars are of equal length. The cross link is half the length between the fixed pivots with the tracer in its center. The center and extreme points are in a straight line.

636. THE DEAD CENTER PROBLEM. Two cranks and treadles with the driven crank at an angle hold the treadle crank in

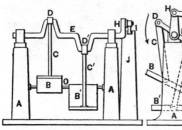

position for starting. J, spring, I, connecting rod, to hold the crank in position for starting.

637. Side view of cranks and treadle connections.

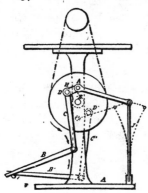

638. THE DEAD CENTER PROBLEM. The crank pin is held off the center by the spring J, the tension of which always pushes the crank pin off the center. Shows the action of a single treadle in the two extreme positions.

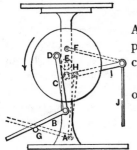

639. THE DEAD CENTER PROBLEM. A supplementary crank set at an angle with the pedal crank and a spring J, to bring the pedal crank to the proper position for starting.

The dotted lines show the opposite position of a center hung treadle.

640. CRANK SUBSTITUTE. The shaft to be driven has recesses in which are pawls or friction devices, two rings being placed on the sleeve having internal ratchets when pawls are used, while bands are connected to the rings and to a frame so that when the frame is moved downward one of the rings on the sleeve will move the balance wheel, while as the frame moves upward the other ring drives the balance wheel, the opposite pawls or friction device slipping over their respective rings alternately with the contrary movements.

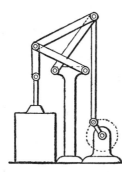

641. SHORT-RANGE WALKING BEAM. By the interlocking linkage the cylinder and crank can be brought close together.

None of the motions in this linkage are parallel. The piston rod is the guide to the last link.

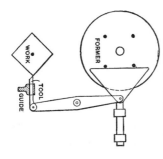

642. TURNING A SQUARE BY CIRCULAR MOTION. A device one hundred and fifty years old. Not an economical device for square work, but applicable for irregular and fluted work. Possibly the original idea of the rose lathe.

643. DOUBLE-LINK UNIVERSAL JOINT. This arrangement allows of a large deviating angle in the line of shafting. The pins have each a clear way through the swivel blocks.

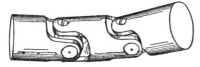

644. CHANGE SPEED PULLEYS. Lazy-tongs type. The device comprises two pulleys, A and B, the rims of which are made in sec-

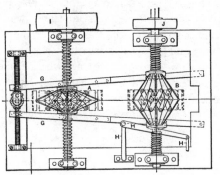

tions so that their diameters can be varied. By turning the crank C the diameter of A is altered, while that of B changes under pressure of the helical springs surrounding its axle, thus keeping the tension of the belt practically constant.

The rim sections or shoes of these pulleys are supported upon a felly, or framework formed of two lazy tongs joined at their summits and pivoted together at the middle of their branches so as to form a series of equal diamonds which must all elongate or flatten simultaneously.

645. MULTIPLE-SHAFT DRIVING DEVICE. The four crank pins are pivoted on an oscillating and sliding sleeve on a central post,

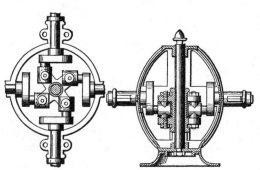

as shown in the plan and vertical section. Either shaft may be the driver. All the shafts must be at right angles with each other, and in the same plane for perfect action.

646. Vertical section showing central sliding post.

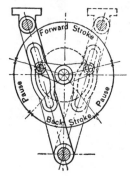

647. RECIPROCATING WITH STOP MOTION. A swing lever operated by a crank may have two stops in each revolution by the opposite curves in the slot of the lever, which are circular, having their radii to correspond with the distance from the crank center to the outside of the crank pin.

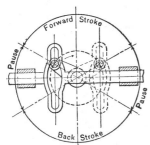

648. RECIPROCATING MOTION with a stop at each stroke from uniform crank motion. The crank pin follows the opposite curves in slot at each half revolution. Rebounding at the wide part of the slots is opposed by buffer springs.

649. RECIPROCATING INTO ROTARY MOTION WITHOUT DEAD CENTERS. The cross-head B, with the peculiar slot C, and offset at D carries the roller crank pin over the center.

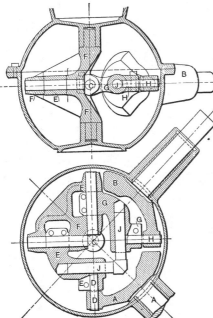

650. RIGHT-ANGLE COUPLING for revolving shafting.

A, driving shaft and crank. B, driven shaft crank. C, point of intersection of shaft centers. D, driving crank pin, pointing to the center C. E, connecting arm. F, oscillating piece with pins pointing to the center C. G, connecting arm to crank pin of driven shaft at H. J, a right-angle motion piece to prevent the driven shaft sticking on the dead center. It has two motions in each arm.

651. Vertical section of the oscillating piece with swivel joints in the shell.

652. REVERSIBLE FRICTION RATCHET. Motion is transmitted to the shaft by a set of friction rolls and the hardened steel block

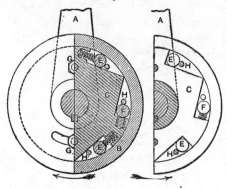

C, to which the shaft is keyed. As the casing is oscillated in one direction or the other, one set of the steel rollers, E, E, or F, F, becomes bound between the block and the casing and causes them to revolve together. As soon as the direction of rotation of the casing is reversed the rolls are freed from their contact and the casing is moved backward independent of the block.

In order to hold the other set inoperative, a cover plate is placed over the face of the ratchet block and fastened to it by two bolts, G, G. At the points where these bolts pass through the plate are two grooves which allow the plate to turn.

This plate is fitted with six retaining pins, H, H, H. When the plate is moved so that the bolts are at one side of the slot, these pins hold one set of rollers out of action as shown.

653. Half section showing one set of rollers held back by the pins and plate.

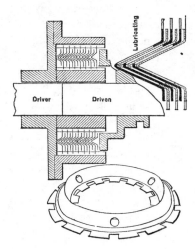

654. FRICTION-PLATE CLUTCH. In this model the plates are pressed into V-shaped rings with perforations for lubrication. The V shape allows of a great friction with light pressure on the clutch lever.

Alternate V plates are fixed to outer shell by their mortised edges, and the intervening plates to the inner hub in the same manner. The perforations for lubricating are shown in the lower section of the cut, No. 655.

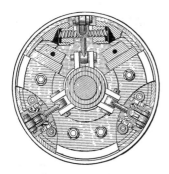

656. FRICTION CLUTCH.   Brown type.   The usual sliding sleeve on the shaft and connection to an arm on a right and left double-thread screw, which expands the friction blocks and so clutches the inner face of the pulley rim.

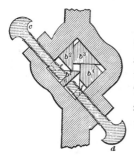

657. EXPANDING WRENCH OR CHUCK. One of the triangular jaws is recessed to form an abutment for the adjusting screws c, d, and two other jaws are slotted to pass over the screws. The square can be varied in size to fit various sizes of tap shanks or drill shanks when the device is used as a chuck.

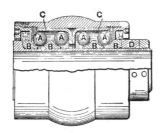

658. MULTIPLE BALL BEARINGS, for vehicles.   The four rings of balls A, A, A, A, are held in place by the ring cones, B, B, B, B, and the whole held in place by the nut and check nut D. C, C, channel sleeves that give the balls a three point bearing.

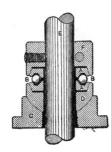

659. SHAFT - THRUST BALL BEARINGS on a vertical shaft.   A, A, grooved rings with outside conical bearings.   D, a spherical bearing collar resting on the foot flange C.   F, retaining collar.

The balls have four point bearings.

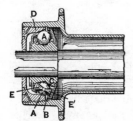

660. BICYCLE BALL BEARING, with hourglass separating rollers. The balls have three points of pressure contact, two on the cone and one on the cup.

The separating rollers are carried by a guide ring frame.

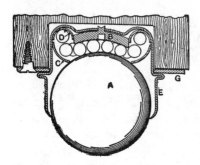

661. BALL - BEARING CAS-TOR. The rolling sphere A is held in position by the sheet met-al case E. About 40 small balls are arranged to circulate under the bearing plate B, guided and held in place by the case C. The balls traverse around the annular space D.

662. SPRING MOTOR. A series of coiled springs and drums arranged side by side on a shaft, and combined together and with the

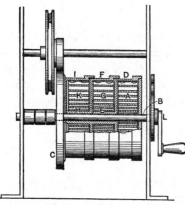

winding-up mechanism and trans-mitting mechanism in such manner as to constitute in effect one spring of great length but in separate coils, which gives much better results in practice than a single spring of the same length in a single coil will.

The first spring A is attached at the inner end of the coil to the wind-ing-up shaft B, which also serves for mounting the spring drums and the transmitting wheel C. At the outer end of said coil this spring, A, is attached to the hollow drum D mounted loosely on the shaft. This drum has a central hub, E, extending along the shaft B within the second drum F, and the spring G in said drum is attached to said hub at its inner end, the outer end being attached to the drum F. This drum F also has a hub H, extending into drum I, and the spring K therein is at-tached to it and to the drum as the others are.

663. SPRING MOTOR.   A pair of shafts arranged parallel to each other, and geared together so that one turns faster than the other, and a

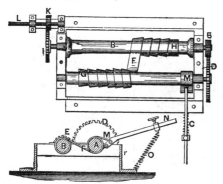

long India-rubber or other elastic band or cord wound or coiled on the shaft which moves slowest, then attached to the other and wound on to it from the first in a way to stretch the band through its whole length, a n d so that, when the shafts are released, motion will be imparted to them by the spring, which will wind back on to the first shaft.

A is one of the shafts and B the other.   They are arranged on a frame, and geared together at one end by the large wheel D and the small one E.

F is the India-rubber belt.   It is fastened at G to the shaft, and wound spirally thereon, as shown, the coil extending from end to end of the shaft; then the other end is attached to the shaft B at H, and, the shafts being turned by hand, the belt will be wound off from A and on to B, and at the same time stretched as much as is due to the difference in speed of the shafts.   As the shafts revolve in opposite directions, the band winds from the top of one to the bottom of the other.

664. Section of shafts, gear and spring brake.

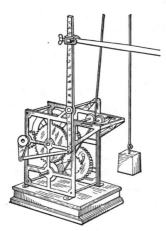

665. WEIGHT-DRIVEN MOTOR. A gear train and winding drum for a rope ; a ratchet wheel and pawl for winding up the weight.   A fly wheel, shaft, and crank gives a reciprocating motion to a lever or for any purpose of motion.

666. SPRING MOTOR. With continuous motion while winding. A, spring. B, drum attached to driving gear C. E, ratchet, fast on shaft.

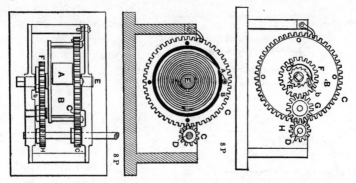

F, gear loose on shaft and carrying pawl *b*. G, idler gear between E and D. Used for running sewing machines.

667. Section showing spring and driving gear.

668. Plan with driving and winding gear, which does not stop the motion while winding.

669. WEIGHT-DRIVEN MOTOR. The power is furnished by the two weights shown, one on each side, ropes from which are carried

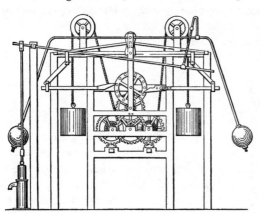

to and are wound around two drums, which form part of clockwork mechanism, with pallet wheel and escapement. Immediately below the wheels attached to the power drums are pinions with square-headed shafts, on which handles can be placed, and which are used to wind up the weights. The frame which carries the two pawls engaging the 'scape wheel is pivoted directly in a vertical line above the axle of the 'scape wheel, and as tooth after tooth of the wheel passes a pawl the frame rocks like the walking beam of a steam engine.

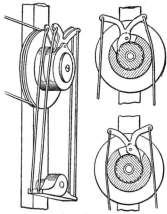

670. SWING MOTOR. A wheel has a hub with two sets of ratchet teeth standing in opposite directions, as shown, collars fitting loosely on the hub over the ratchet teeth, and pawls fulcrumed on the collars to engage the teeth, while a lever mounted to swing on a stud is connected by belts to arms extending outward from the pawls. The wheel with its hub is held in place on the shaft by a washer, which also serves to hold the collars in place.

671. Section showing ratchet and pawl for forward motion.

672. Section of ratchet and pawl set for backward motion.

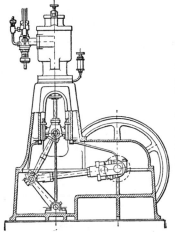

673. AMMONIA COMPRESSOR. Two strokes of a single-acting piston to each revolution of the crank by the double-acting toggle. In this design, the action of the toggle compensates the difference of pressure in the steam and ammonia cylinder. National Refrigerator Co. type.

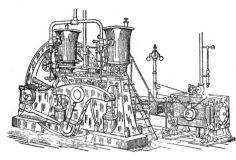

674. AMMONIA COMPRESSOR. Illustrates the T, crank movement for operating a duplex compressor with single-acting pistons. Cylinders are overhead with water jackets.

675. COIN-IN-THE-SLOT GAS METER. A coin dropped in the slot falls on the lever L, and by depressing it locks the outside handle

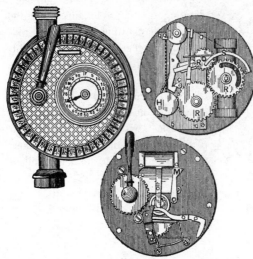

to the plug of the gas cock. The opening of the cock by the handle sets a spring and winds up a small clock movement, setting it in motion. A small cam, C, is set in motion against the lever D, and releases the weight H, which has been lifted by the sector S and clock train at the moment when the measure of gas allowance is made.

676. Part of the clock train, releasing lever and driving weight.

677. Slot passage to coin lever, handle and winding gear.

678. SPIRAL FLUTING LATHE. The baluster is fed endways against a lateral tool, being rotated on its axis at such a rate as shall im-

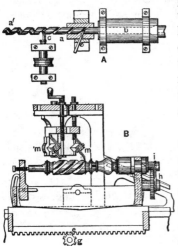

part the number of turns or part of a turn to the foot in length. Of this kind is A, in which the piece a to be cut is moved longitudinally through the holder b, and at the same time is rotated so that the tool c in its revolutions may cut the spiral groove shown at a'.

679. B is a fluting lathe in which a pair of cutters revolve in a plane oblique with the line of motion of the baluster. The latter is moved longitudinally by the rack and pinion e g, and rotated by the wheel and pinion h i; the cutters m m rotating in parallel planes cut two grooves at once.

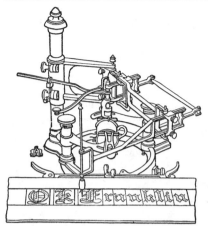

680. PANTOGRAPHIC EN-GRAVING MACHINE. A cup or any article to be engraved is held in clamps in the central part of the machine and under the cutting tool. The stile or tracer is at the long arm of the pantograph and follows the pattern figures or letters, while the engraving cutter is pressed upon the work by a lever.

681. GEOMETRICAL BORING AND ROUTING CHUCK. By means of a set of cam gears within the square box and adjusting

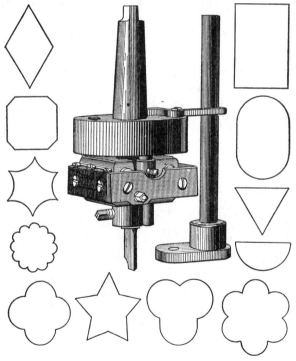

screws, a variety of shaped holes, recesses, or indented figures may be cut by variously shaped cutters. The lever against the guide bar checks the revolution of the geometric cams.

682. The figures are the curves and forms produced by the chuck.

683. A ROSE LATHE or engraving machine. The principal feat-

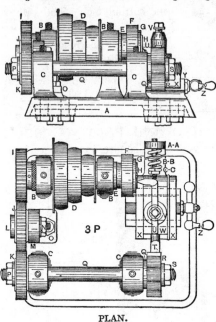

PLAN.

ures are well shown in the cuts, and the specimen cut shows a few of the designs that such a machine is capable of producing. The sectional head contains the clamp device for holding the work. The tool U may have from one to four points like a chasing tool to vary the design of the work. H is the work in the chuck and R the cam or rose plate. The follower stud T is mounted on the tool post slide and held against the rose plate by a spring. The relative sizes of the gears, I, J, K, may be varied for a great variety of figures.

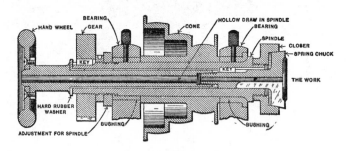

SECTION OF SPINDLE.

684. Plan.

685. Section of spindle.

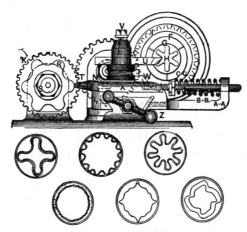

SECTION AND EXAMPLES.

686. Tool post, rose wheel and cutting tool at the work at H, on the face plate of the lathe.

687. Examples of curved figures made by different forms of rose wheels.

688. PLANETARIUMS. In the lower planetarium the globe representing the sun is supported on a central shaft, around which are ar-

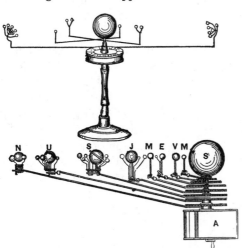

ranged a series of sleeves, corresponding in number to the planets of the solar system. The shaft supporting the sun is caused to rotate in a time relatively corresponding to the diurnal revolution of that luminary, and the sleeves which carry the tubes supporting the planets are also revolved in times proportionate to their revolutions around the sun by wheels meshing with gears on a shaft within the case A, and provided with an exterior crank by which it is turned. The diurnal revolutions of the planets are caused by bevel-gearing on the sleeves and on rods within the tubular arms above mentioned, which rods also carry on their ends gears for causing the revolution of the satellites around their primaries.

See Nos. 984 to 992 first volume of mechanical movements for details of planetary gear trains.

689. Planetarium of the solar system.

690. THE PHENAKISTOSCOPE. This instrument, which, like the *thaumatrope* and *zeotrope*, depends upon the persistence of visual im-

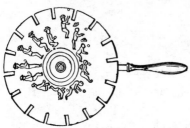

pressions, consists of a circular disk on which a row of figures are painted in a series of attitudes such as would be consecutively attained in the progress of an action ; for example, leaping, walking, swimming, etc. The effect is to produce the appearance of actual motion. The disk is placed on a handle and rotated by the finger on a nut. It is held in front of the observer, the face of the toy toward a looking-glass, and the figures are viewed through the slits.

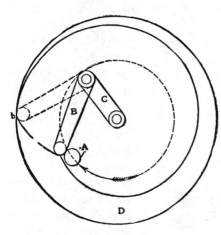

690A. LOST MOTION CONNECTION. The pin *A* moves as shown by the arrow, pushing lever *B* ahead of it. One end of this lever is pivoted to *C* and the other end rests against the cam *D*. While *B* is moving from *a* to *b*, *C* stands still. The time thus lost is made up during the rest of the revolution. Thus *C* revolves around the center every time *A* does.

SECTION XVI.

HOROLOGICAL, TIME DEVICES, ETC.

Section XVI.

HOROLOGICAL, TIME DEVICES, ETC.

691. ELECTRIC PENDULUM.   P P is the pendulum, W a weight, mounted on a lever, W C A.   W C A can move about a center, C, and

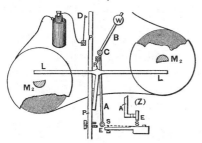

is at present prevented from turning by the catch, S S.   When P P swings to the right, the lower screw in P P passes under E (see side view Z) and frees W C A. W C A, under the weight of W, propels the pendulum to the left till stopped by a banking, B.   P P moves on and makes contact with D, whereupon a current passes, M_1 M_2 become magnetized, and attract L L, the vertical arm of which lifts W C A over the catch, S S, again. When P P leaves D the current ceases, and L L is carried back to its old position by the action of the spring R.

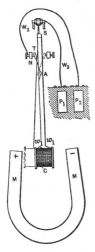

692. ELECTRIC PENDULUM C is a bob in the form of an electro-magnet vibrating between the poles of a permanent magnet.   T and N, reversing switches operated by the motion of the pendulum.   S, direct connection to the ground batteries P_1 P_2.   The current is reversed as the pendulum bob nears each pole of the permanent magnet.

Magnetic clocks are thus made continuous in operation by a simple gear train and dent escapement, as shown in other figures.

693. ELECTRIC CLOCK CONTROLLER. The pendulum at the right hand is of the controlling clock, and the central pendulum that of a controlled clock; the

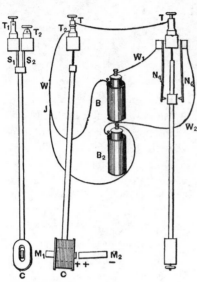

pendulum at the left is a side view of the central one. C, the bob, is a hollow coil of insulated wire, and swings over two magnets, $M_1 M_2$, which have their similar poles facing each other. The ends of the wire forming C are carried up the pendulum, pass respectively through $S_1 S_2$, and terminate in $T_1 T_2$. T_1 is joined to T, which crowns the pendulum of the controlling clock, and T_2 is in connection with both $N_1 N_2$, the contact springs of the same. Both $N_1 N_2$ have their respective batteries, $B_1 B_2$, but with opposite poles toward J; so that if C is magnetized in one direction by one swing of the pendulum, it will be magnetized in the opposite by the other, thus making a synchronal beat by the controlled clock.

694. REPEATING CLOCK. M, air-bulb tube with piston for moving the stop lever K. O, a push-button switch for operating the

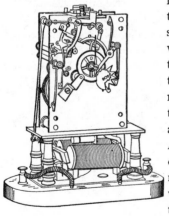

lever K by the electro-magnet. At rest, the different parts are in the position shown in the diagram (No. 695), and the wheelwork is arrested by a snug fixed to the piece H. Upon the piece, A, are fixed two pins, A' and A", which are so arranged that after lifting the detent, G, the latter may drop just at the moment at which the hour hand is upon 12 or 6. As soon as it is raised, the detent, G, carries along with it the stop H. At this moment there occurs the first start of the wheelwork, the detent, G, falls and sets the wheelwork free. The piece, H, remains raised (the arm, H', engaging with the teeth of the rack) and permits the wheelwork to continue its revolution. The rack is raised

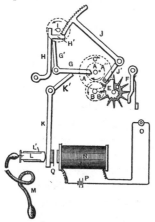

tooth by tooth by the click, I, fixed up-on the second wheel. To every revolution of the latter there corresponds one blow struck upon the bell of the clock. As soon as the rack is lifted high enough, the piece, H, falls to its position of arrest in stopping the wheelwork. In order to cause one stroke only to be given at the half hour, the pin, A'', is fixed upon a smaller diameter, so as to raise the pieces, G and H, sufficiently to permit of the first start, but not enough to cause the rack to fall. The wheel-work is therefore arrested as soon as the second wheel has made one revolution. Thus by an electric push-button or a compressed-air piston a clock may be made to ring the nearest hour or half-hour at any time at night.

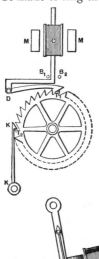

696. ESCAPEMENT WITH ELECTRIC PENDULUM. M, M, permanent horseshoe magnet. $B_1 B_2$ stops to limit the motion of the detent D, D. K, K, stop click to hold the tooth. For operating circuit clocks by current sent from a central clock beat, through the electro-magnet, C.

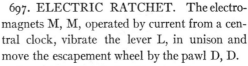

697. ELECTRIC RATCHET. The electro-magnets M, M, operated by current from a central clock, vibrate the lever L, in unison and move the escapement wheel by the pawl D, D.

A very simple device for operating the es-capement of a secondary clock from a central station.

698. SOLAR AND SIDEREAL CLOCK. Firmly secured on a solid base of metal are two regulators, each having a one-second mercurial

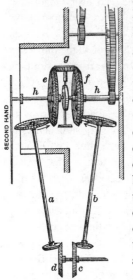

pendulum. One of the pendulums is regulated to mean solar time and the other to sidereal time, the dial of the latter being divided into 24 hours and that of the former into 12 hours. The escape-wheel shaft of each clock is long enough to reach out through the dial plate, and on the outer part is fitted, with a slight friction, a sleeve. On the inner ends of these sleeves are the beveled wheels, c d, of 90 teeth each, and their outer ends carry pointers indicating seconds on the dial plates. Engaging with these v heels are beveled pinions, of 30 teeth each, mounted on the lower ends of the long shafts, a b, which are carried up at an angle of about 45 degrees and connected with a differential motion controlling the works and hands of a larger dial placed above the two others. This peculiar motion is constructed of a light shaft, h, on which is fastened at right angles a crosspiece, on one end of which is mounted the wheel, g. On the shaft, h, and engaging with the wheel, g, are two

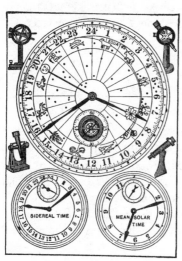

699.

larger wheels, e f, of 90 teeth each; these wheels are cut on both sides, as shown. Engaging with these wheels are wheels of 60 teeth each, fastened on the upper ends of the shafts a b. It will be seen that both clocks are directly connected with the differential motion, and also that as long as the wheels, e f, which turn in opposite directions, are driven at the same speed, the wheel, g, will simply roll on its pivot without altering .its position or that of the shaft h. But assuming that the wheel, f, revolves twice around while the wheel, e, revolves once, then the wheel, g, will necessarily follow f, and in proportion to the speed of the two wheels, *e f;* but as these wheels move in opposite directions, it consequently follows that one-half the difference

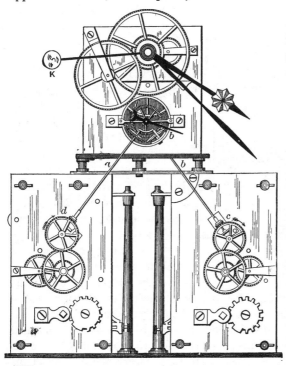

700.

in the rates is lost, or instead of making a complete revolution—the difference between 1 and 2—it has only recorded half a revolution.

Now, to compensate for this error—in other words, to regain the half revolution lost—the wheels on the upper ends of the shafts, *a b*, have 60 teeth each, and the pinions at the lower ends have 30 teeth each; and as the driving wheels, *c d*, having 90 teeth each, are connected through the pinions, shafts *a b*, and upper wheels with the wheels, *e f*, also of 90 teeth, the differential motion will be compensated.

Now, as the clock marking sidereal time gains at the rate of about 4 minutes in 24 hours, or 10 seconds in 1 hour, and as 10 seconds is one-sixth of a minute, it will take 6 hours to complete one revolution of the hand on the differential motion, which is the period of 1 minute in right ascension; 15 days 6 hours is 1 hour, and 1 year is 24 hours in the same measure. The hour hand on the large dial, therefore, represents the sun's apparent yearly motion among the stars.

701. NOVEL CLOCK.   The novelty of the clock consists principally in the escapement.   Beneath the main mechanism is placed a

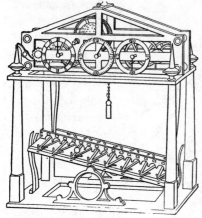

tilting table pivoted upon studs projecting from the center of its long sides, so that it is free to have a seesaw movement.   Upon the upper surface of the table is formed a zigzag groove in which travels a small steel ball.   The path is made up of sixteen divisions, so that the ball, starting at the elevated end of the groove, passes across the table, forward and back, until it reaches the lower end, which is then elevated to enable the ball to run back to the starting point, which is again raised, and so on.

Attached to one end of the table is a rod leading upward to an arm placed at right angles on the end of a shaft driven in the usual way. When the ball reaches the depressed end of the table, it strikes a spring which releases a catch holding the shaft, which is thereby permitted to make a half turn, and its arm is correspondingly moved to raise or depress that end of the table to which the connecting rod is attached.   The ball then runs down the table, strikes a similarly arranged spring at the opposite end, when the movements are repeated and the position of the table again reversed.   It takes fifteen seconds for the ball to travel from one to the other end of the table.

702. ELECTRICAL CORRECTION OF CLOCKS.   For a clock

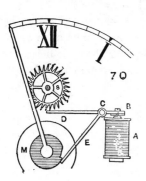

that gains some second or two per hour. Fifteen seconds before each hour the lever, D B, is attracted by the electro-magnet, A, and a pin in the arm, D, would thereupon enter and catch a tooth of the escape wheel, did the disk, M, allow the other arm of the lever, E, to move.   When the hand reaches the hour, E falls, then D catches S and holds it till the cessation of the current at the sixtieth second of the governing clock.

703. LONG-DISTANCE TELEGRAPH-CLOCK CORREC-
TION. Generally the use of a long telegraphic wire can only be com-

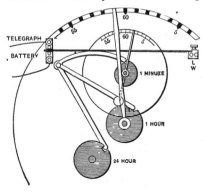

manded for a few minutes daily. The cut shows a very suitable arrangement to be adopted when this is the case. By means of the 24-hour disk the line wire is held in communication with the telegraph office until a few minutes before the clock current is going to be dispatched. The notch in the 24-hour disk will at last allow the system of levers to fall, but then the one-hour disk supports them until about one minute before the clock current is coming; so that, till then, the line is being used for messages. The line wire has not been allowed to fall into circuit with the battery wire ; this is still prevented by the one-minute disk. At the sixtieth second precisely, the one-minute disk allows the line wire to join the battery wire, and out goes the clock current. Some seconds afterward the one-hour disk lifts the line wire back into communication with the telegraph office, where it stays for another 24 hours.

704. FLYING-PENDULUM CLOCK. The central vertical spindle tends to revolve continuously by virtue of its connection with

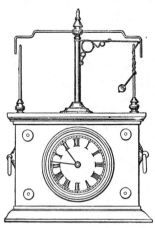

the driving gear of the clock, but when the arm which it carries swings halfway round, the little spherical weight, suspended from it by a thread, is thrown outward by centrifugal action ; and when the thread touches one of the fixed vertical wires at the side of the clock, the momentum of the spherical weight causes it to wind the thread around the vertical wire and stop the arm and spindle. As soon as the thread is wound upon the spindle, the spherical weight unwinds it by its own gravity, and in so doing receives enough momentum to rewind the thread and still prevent the spindle from revolving. Then the thread winds and unwinds once more, when the arm is released, and makes a half revolution, when the thread is wound on the other vertical wire, and the operation just described is repeated.

705. SELF-WINDING, SYNCHRONIZING CLOCK. O, P, electric motors operated by a local battery. Clocks used in the syn-

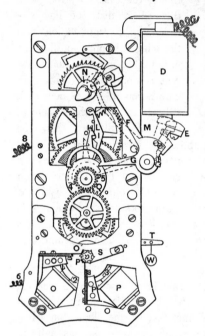

chronizing circuit are provided with the synchronizing magnet, D, and with mechanism associated with its armature lever and the clock movement. On the minute-hand arbor is mounted a disk, Q, provided with two projections, 4, 5, and the second-hand arbor is provided with a heart-shaped cam. The armature, E, is rigidly attached to the levers, F, G, so that they move whenever the magnet, D, is energized. The lever, F, is adapted to engage the heart-shaped cam on the second-hand arbor and bring it to XII, and the lever, G, is furnished with a curved end having fingers for engaging the projections, 4, 5, on the minute disk, thus turning the minute-hand arbor, bringing the minute hand to XII.   A latch, L, pivoted to the clock frame, is provided with a pin, I, arranged to drop under the hook, H, carried by the lever, G, so as to prevent any action of the synchronizing levers except at the hour.   A pin in a disk mounted on the cannon socket unlocks the latch, L, about fifty seconds before the hour, and closes it again about fifty seconds after the signal. This arrangement prevents any accidental cross on the synchronizing line from disturbing the hands during the hour.

SECTION XVII.

MINING DEVICES AND APPLIANCES.

Section XVII.

MINING DEVICES AND APPLIANCES.

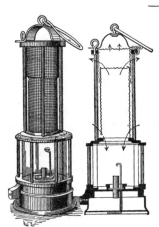

706. MINING LAMP. Clanny type. A glass takes the place of the lower part of the wire gauze in the Davy lamp and thus gives a clear light from the flame. The air enters through the lower part of the wire gauze chimney, as shown by the arrows in the section. English.

An improvement on the Davy lamp.

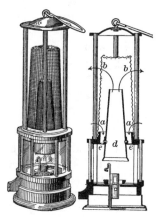

707. MINING LAMP. Mueseler type. A strong glass cylinder around the flame section. A central metallic chimney *d*, arranged to separate the incoming air at *a* through the wire gauze, while the products of combustion pass up the central chimney.

By this arrangement the flame is fed with purer air than when the central chimney is not used.

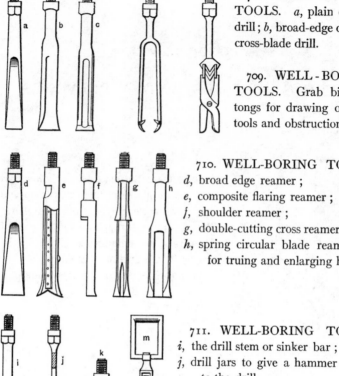

708. WELL - BORING TOOLS. *a*, plain driving drill; *b*, broad-edge drill; *c*, cross-blade drill.

709. WELL - BORING TOOLS. Grab bits and tongs for drawing out lost tools and obstructions.

710. WELL-BORING TOOLS.
d, broad edge reamer ;
e, composite flaring reamer ;
f, shoulder reamer ;
g, double-cutting cross reamer ;
h, spring circular blade reamer, all for truing and enlarging holes.

711. WELL-BORING TOOLS.
i, the drill stem or sinker bar ;
j, drill jars to give a hammer action to the drill ;
k, short sinker bar above the jars;
l, temper screw to adjust the length of rope for properly operating the jars and drill ;
m, clevis or walking-beam strap.

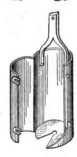

712. WELL-BORING TOOLS. Sand auger. The bottom is a spiral bit for catching the sand by turning the auger. A door on one side for discharging the sand.

713. WELL-BORING TOOLS. Portable power rig and drill beam.

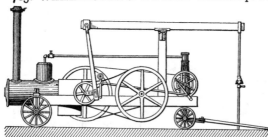

A frame tower is also used a b o v e the drill hole for a sheave over which the drill rope is passed a n d to a winch driven by the engine.

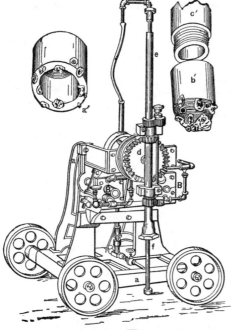

714. PROSPECTING DIAMOND DRILL. A hollow drill bar, f, slides in the revolving hollow shaft, e, driven by the bevel gear, d, from a motor. Water is forced through the drill rod by a pump. The drill is tipped with a steel ring studded with black diamonds, so that the drill makes an annular cut by which c o r e s may be taken out for examination. The tailings of the drill are washed to the surface by the force of the water jet. b is a solid diamond-set drill; a, a core drill.

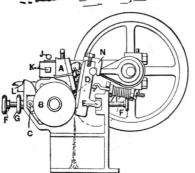

715. ASSAY ORE CRUSHER. A combination of jaws and finishing roller, adjustable to crush the ore sample to a uniform size.

F, F, adjusting screws for roller. C, friction thrust rollers. B, finishing roller.

716. ORE ROASTING FURNACE.   Section of the Pearce turret furnace in which a circular oven is heated by outside fires. With rabbles

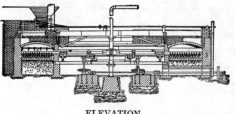

ELEVATION.

on arms from a central revolving shaft constantly stir the ore and move it forward around the hearth to a discharge hopper. The plan and vertical section show much of the detail of this class of roasting furnace.   The rabble arms to which the plows are attached are hollow, and are cooled by air or water. The furnace is entirely automatic, the ore being dropped

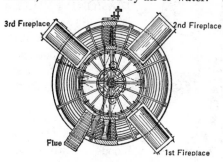

PLAN.

from the ore hopper at intervals and carried into the hearth of the furnace. After traveling around the hearth it is discharged into a chute, which delivers it to either a car or a cooling apparatus, from which it passes to the elevators. When air is used to cool the rabble arms, the hot air may be delivered to the hearth at any point desired, by means of a simple automatic device.

717. Plan of turret roasting furnace with fireplaces and flue.

718. ORE ROASTING FURNACE.   Two-hearth furnace of the Pearce turret type.   The width of hearth is 6, 7, and 8 feet, and the fur-

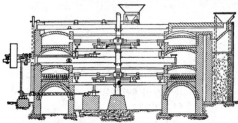

nace may be built so that the top arch acts as a drying hearth.   The number of fire boxes varies from two to three, according to the ore, the process by which it is to be subsequently treated, and the fuel.   When petroleum residuum is the fuel, the projecting fire boxes are omitted and a combustion chamber is built directly over the hearth, through which the oil burners project and distribute the flame over the whole width of the hearth.

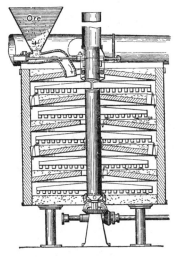

719. ORE ROASTING FUR-NACE. The ore fed from the hopper at the top is drawn alternately inward and outward on the partition hearths of the roaster by revolving arms with blades inclined to draw each way on alternate floors. The hot gases enter from the flue at the top and are discharged with the ore at the bottom of the roaster. Herreshoff type.

720. ORE ROASTING FURNACE. Straight line type. Ropp. The ore enters from a spreading hopper at one end and is drawn by

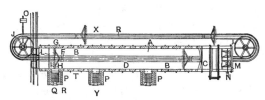

rabbles with inclined teeth for alternate turning over of the ore during its passage along the furnace bed to the discharge pit. The rabbles are attached to a chain carrier and returned on the outside of the furnace. The furnace grates are on the outside, distributed for equalizing the heat which passes off through a chimney at the feed hopper end.

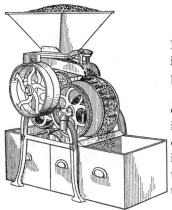

721. MAGNETIC METAL SEPA-RATOR. For separating iron turnings, filings, and chips from brass composition or other non-magnetic material.

The drum is composed of the faces of a large number of magnets. The iron adheres to the magnets and is carried around the cylinder to a revolving brush, while the non-magnetic material drops from the front of the drum to a box.

722. MAGNETIC SEPARATOR for separating iron from brass turnings, small scrap, chips, and drillings. A rod connected to the

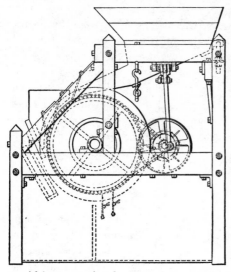

hopper swings it back and forth when the separator is in operation so as to distribute the metal on the surface of the drum. An adjustable grate is provided so as to regulate the flow of metal from the hopper. The length of the swing of the oscillating mechanism is adjustable.

Current is supplied to the electro-magnets within the drum through collector rings and carbon brushes. The flanges at the edges of the drum keep the metal on the surface. The metal is stirred to aid in separation by two wires supported from the frame and extending across in front of the cylindrical surface. The brush wheel which removes the iron from the cylinder has radial strips of sole leather secured between wooden blocks. This wheel rotates in the same direction as the drum, but at a much higher speed. The bins into which the separated metals are dropped are situated within the frame and open at opposite ends of the separator so as to prevent the metals from becoming mixed in handling.

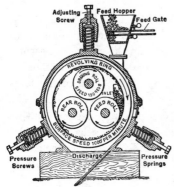

723. QUARTZ PULVERIZER. Kent type. A revolving cylinder ring inclosing three revolving rollers traveling with the ring. Quartz from the crusher is fed through the hopper at one side of the cylinder, and the finely pulverized material discharged at the other side.

724. ORE WASHING TOWER. The rock is delivered through a hopper to a vertical conductor, which has a series of inclined plates or

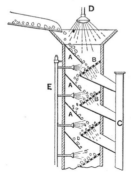

aprons, A, and opposite perforated plates, B, the rock falling first upon one and then another of these plates in its passage downward through the conductor. Over the conductor is a rose nozzle, D, which showers water upon the rock, and opposite each of the perforated plates are jets supplied from a stand pipe, E, the water thus sprinkled on the broken rock passing down the conveyor carrying off the refuse matter through the chute C. The number of the plates and their inclination and arrangement may be varied according to the nature of the material to be treated.

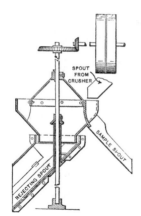

725. AUTOMATIC ORE SAMPLER. The revolving hopper has one or more small aprons in its periphery which divert a small portion of the ore passing through the sampler to the sample spout. The ore is still further crushed and passed through a second and third sampler, so as to obtain a fair sample to the one hundredth or more part.

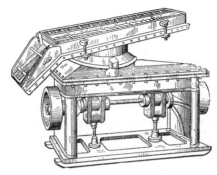

726. PNEUMATIC CON-CENTRATOR. The vibration of the vanner and concentration of gold sands is done by short, quick strokes of air pumps or bellows beneath the table which are operated by the crank shaft.

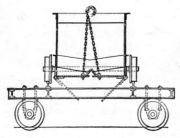

727. ORE CAR ON A TRANSFER TRUCK. The transfer truck has an open platform through which the ore is dumped by dropping the bottom of the ore car, which is held by chains and a windlass.

728. DRY PLACER GOLD SEPARATOR. Edison type. The revolving roller, *b*, discharges the gravel from the hopper, *a*, upon the shelf,

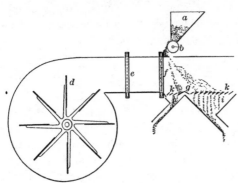

c, from which it falls into the air blast created by the centrifugal fan, *d*, discharging its air through the screens *e* and *f*. The parting board, *g*, divides the heavier portion of the gravel—the gold and iron or black sand, which falls into the chute, *h*, from the lighter portion falling into the tailings chute *i*. The lattice, *k*, *k*, is simply to prevent eddy currents of air going down the chutes *h* and *i*. The end of the air pipe at *k* is open.

By a suitable adjustment of the speed of the fan, the position of the parting board, *g*, and the rate of feed of the gravel concentrates are obtained ; the screens are necessary for equalizing the velocity of the air blast.

729. DRY GOLD MINING MACHINE. The hand crank wheel, the vibrator, and the blower for blowing off the dust and sand from the riffle table are the leading features of this novel dry placer machine. Gold sand is fed to the hopper above and shaken in a thin sheet to the riffle through a blast of air. Air is also blown through the sieve sections pushing the sand forward and holding the gold.

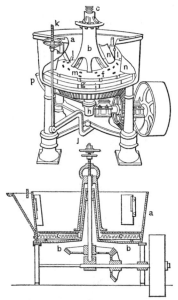

730. GOLD AMALGAMATOR.
a, the circular tank with sluiceways for overflow of the waste slimes.

c, c, revolving stirring arms driven by the gear and center shaft through the conical center standard of the tank. The amalgamated plates or mercury rests on the bottom of the tank.

m, f, perforations in the disk carrying the stirring arms, for equal distribution of the ore slimes upon the mercury bed.

731. Section showing gearing, elevating screw and stirring arms.

732. SHEAVE WHEELS FOR GRAVITY PLANES. The front wheel is made smaller than the rear wheel (usually about 10 inches) so

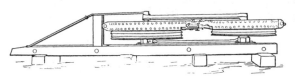

as to allow the rope to lead from the rear wheel to the knuckle sheaves, and permit of additional room at the top. This front wheel has one or more grooves, depending upon the number of cars to the trip and the amount of material to be hauled, and has always one less groove than the larger or rear wheel, the sheaves being placed tandem. The rear wheel is made with two or more grooves, to allow the rope to be placed on the sheaves in the form of a figure 8, thus securing considerable contact of the rope on both wheels.

The depth of the grooves in each wheel should be exactly the same, a variation of a small fraction of an inch being detrimental to the life of both rope and sheaves, as if the wheels are not made correctly, the rope passing around them must slip or stretch with each revolution of the wheels.

The brake bands are either lined with cast-iron shoes or maple blocks placed on end, the latter method being usually preferred. Each brake is also provided with a large screw and nuts to take up wear.

733. BRIQUETING MACHINE. Eggette type. Two large rolls

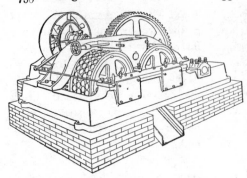

having indentations on their face to correspond with a half-egg shape and made to register, revolve under a close-fitting hopper; the eggettes drop from beneath the rolls upon a conveyor belt, or to a bin through a chute.

734. A BRIQUETING PLANT. A main driving shaft overhead drives by belts, the ore dust mixer, lime tank mixer, briqueting machine, and conveyor

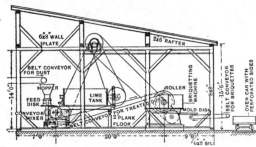

belts. The material used for cementing the briquets varies greatly with the kind of material to be briqueted. For ores, lime is in general use. For fuels, coal tar, resin, pitch, clay, and lignite for anthracite culm. For bituminous culm, lime, clay, and sawdust are used if the coal will not briquet alone under the pressure used.

735. BRIQUETING MACHINE. Plunger type of the H. S. Mould Co., Pittsburg, Pa. The material to be briqueted is mixed

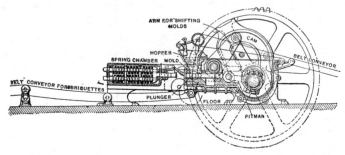

with milk of lime, or any suitable stickative, in a mixer above the machine and fed to the machine hopper, where the compressing plungers press it into molds from which the briquets are ejected by spring plungers at the left on to a conveyor.

736. BRIQUETING MACHINE. A pair of heavy rollers in a circular trough rolls the briquet material into the holes of a revolving mold plate, in which the briquets receive a further pressure and are ejected on to a belt carrier, which deposits them in a truck or bin.

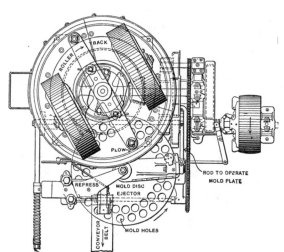

Type of the Chisholm, Boyd & White Co., Chicago, Ill.

737. COAL-WASHING JIG. Coal and slate are washed through a vibrating box or jig, and separated by their difference in gravity. The coal is carried over in a short chute to the coal elevator and the slate is discharged through an adjustable trap to the slate hopper below, from which it is carried away by an elevator belt.

A small vertical engine operates the jig.

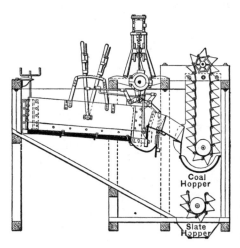

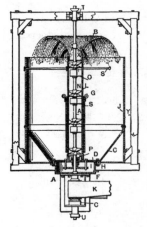

738. PROPELLER PUMP AGITATOR.

A series of propellers on a shaft revolving in a tube draws the solid and fluid material in at the bottom and discharges it in a shower at the top. For agitating oils and ores, as in the cyanide process for gold separating.

In other designs a single propeller is placed at the bottom with the shaft in a step and driven by belt and pulley on the shaft above the tank.

739. COAL-HANDLING PLANT.

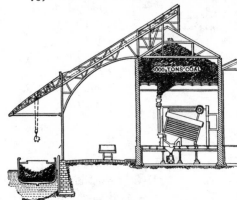

Modern method for conveying coal from boats or cars to overhead lofts for self-feeding to the furnaces of boilers and for dropping the ashes into cars beneath the furnaces and their ready removal. In a long storage loft a track is laid lengthwise, and a car distributes the coal from each hoisting bucket.

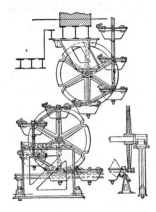

740. METHOD OF CHANGE DIRECTION for conveyor buckets.

Rails support the buckets, which roll on small wheels for horizontal runs. Sprocket wheels take the bucket links for change of direction.

SECTION XVIII.

MILL AND FACTORY APPLIANCES AND TOOLS, ETC.

Section XVIII.

MILL AND FACTORY APPLIANCES AND TOOLS, ETC.

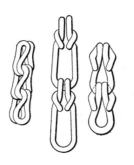

741. MACHINE-MADE CHAINS. Samples of the complex operation of modern machine work. Not only chains of various patterns, but hooks and eyes, and almost every conceivable form of wire work and punch and press work is now done by machinery.

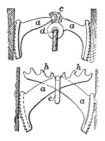

742. SUSPENDING GRIP in a shaft or between timbers. The toothed sectors, *a*, grip the rock or timbers. The upper figure is for a definite sized opening with a pivoted clevis at *c*.

743. The lower figure for a variable sized shaft or opening. The toothed arms, *h, h,* and locking clevis makes a convenient adjustment for any size shaft within its range.

744. UNIVERSAL DOG. Easy to apply to all kinds of work to which it is applicable. Has a great range of size and a good grip.

745. DRILL CHUCK for small drills. A section. By revolving the knurled nut the jaws are moved outward or inward in the converging

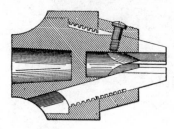

slots in the chuck body as may be desired. The chuck can be operated by hand, and when a very firm grip is desired it may be obtained by the use of a spanner wrench. The chuck may be taken apart readily for cleaning and oiling by removing the three screws in the cap, taking that off and revolving the nut enough to disengage the jaws.

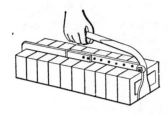

746. BRICK CLAMP. A handy tool for handling brick. It pays by saving the hands.

747. COMBINATION TOOLS. One of the handiest tools

for a farmer or amateur workman is this combination of an anvil, vise, and a drill stock.

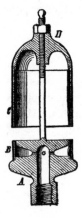

748. EASILY MADE STEAM WHISTLE. A, a brass casting into which the bell stem may be screwed. B, a tube of the same diameter as the bell, soldered to the casting, A, leaving an annular opening $1/100$ of an inch. C, the bell which may be cast or made of the same tubing as at B, with a headpiece, D, soldered in.

749. GASOLINE - HEATED SOLDERING COPPER. The central chamber is half filled with gasoline ; the asbestos wick draws the gasoline toward the needle valve, where the surrounding hot metal vaporizes and discharges the vapor through the small nozzle to be burned

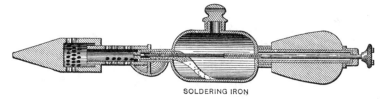

SOLDERING IRON

by the air drawn in through the holes in the tube. The flame impinges on the solid head of the copper and is exhausted through the holes in the headpiece. To start, a little gasoline is poured into the cup under the valve neck and fired for a moment. The valve regulates the amount of flame.

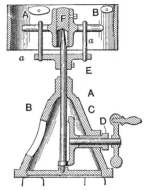

750. P U L L E Y BALANCING MA-CHINE. A pulley poised on a centerpiece, F, is rotated at considerable speed by the arm, E, and studs, a, a. If unbalanced in the plane of revolution, it will wabble, when the high points may be marked with chalk and balance pieces applied as at A and B. The same machine, if set on rubber springs, will show the general unbalanced condition by the vibration of the spindle and the heavy side of the pulley marked with chalk.

751. LUBRICATING DRILL. Holes through the length of the twisted blades carry the oil to the cutting edges and with a constant flow clears the chips by floating them up along the twist grooves. W a t e r may be used for cast iron or com-pressed air fed from a loose socket on the drill holder.

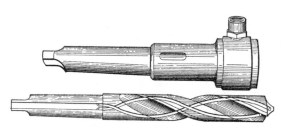

752. Showing holes through thick parts of blade.

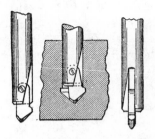

753. EXPANDING DRILL. For enlarging the bottom of drill holes for flush tapping or for Lewis jaws or anchors. The pivoted cutter allows of the cutting of a larger cavity than with the eccentric pointed plain drill.

754. Position of tool when under-cutting.

755. Front view of cutter.

756. TAPER ATTACHMENT TO A LATHE. On the tailstock a place is planed off to serve as a bearing for the guide bar, A, which is

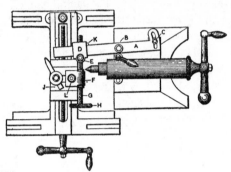

pivoted on the stud, B, and is clamped by the bolt C. Running on this guide bar is the slide, D, provided with a gib, K, to make adjustment for wear. On the under side of the slide is a swivel nut which is fastened to D by the bolt E. Through this nut passes the adjusting screw, G, which serves to connect the guide bar with the cross slide J. The feed block, F, is made to fit into the T-slot in the tool block, being held by a single bolt, L, so that it is the work of but a few moments to remove the attachment when it is not desired to use it.

When the attachment is being used the cross-feed screw is rendered inoperative by dropping the nut or removing the screw altogether. H is a knurled handwheel for operating adjusting screw G.

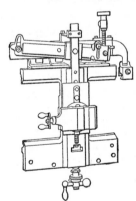

757. TAPER TURNING ATTACHMENT. Bradford type. A taper slide made adjustable for the required taper is fixed to a clamping piece made fast to the back way of a lathe. A slide on the taper member is screwed to the cross slide of the rest ; the nut of the cross screw is cast loose, when the tool follows the angle of the taper bar.

758. CENTERING DEVICE FOR A DRILL PRESS. The device consists of a drill shank, A, made to fit the drill spindle; a centering

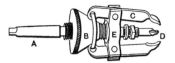

vise with three jaws, two of which are indicated by C, and a chuck to hold the combined drill and reamer D. The centering jaws are closed on the work by screwing down the cone-shaped piece, B, which forces the upper ends of the jaws apart and closes the lower ends on the work. The jaws are retracted by a coiled spring passing through their upper ends. The jaws do not rotate with the centering drill, but remain in a fixed position, being mounted on a sleeve in which the drill shank turns.

759. BORING ELLIPTIC CYLINDERS. B D is a boring bar swung on centers on a lathe. A C is an arm holding at its end C a boring

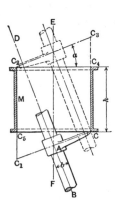

tool. When the bar is rotated the point C describes a circle. A casting, M, which is to be bored elliptic-ally, is secured rigidly to a carriage capable of being moved in the direction of E F, which is the longi-tudinal axis of M and passes through A. Conse-quently, if the boring bar is rotated on its centers and the work fed gradually, in the ordinary way, along axis E F, the cutting tool, at each revolution, describes a circle; but, because of the inclination of axis E F, when the tool occupies the position shown in dotted lines, the boring will be achieved and the end view of the casting will show a perfect elliptical bore.

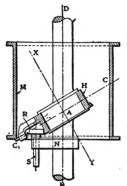

760. BORING ELLIPTIC CYLINDERS. For boring elliptic cylinders of considerable length the cylinder may move forward parallel with the fixed boring bar on which a fixed spool, A, is set at the angle required by the elliptic pro-portions of the cylinder, and upon which a ring, H, and tool holder revolve by a bevel gear and pinion, R, driven by the side shaft S. The tool, C, although cutting in a circular path, by its angular direction produces an elliptic surface due to the angular plane of motion.

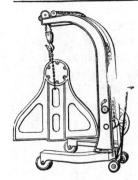

761. C R A N E   T R U C K. One of the handy things in a shop or warehouse. With windlass chain and block two tons may be lifted and easily wheeled over the floor.

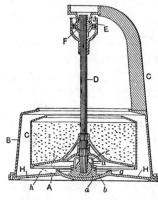

762. CENTRIFUGAL S E P A R A - TOR. The perforated basket and load are hung on the spindle. E is a cup-shaped pulley, inside of which is a ball-socket journal box. The step bearing, b, a, is also a ball socket in a spherical foot, bearing upon a spherical base with a flange on its rim to limit the eccentric swing of the spindle to accommodate the center of gravity for an unequally balanced load.

763. BLACKSMITH'S HELPER. The hammer handle is pivoted at B to the head of a vertical shaft, C, that is fitted in a socket. The

lower end of the shaft has a step in the lever, F, which is pivoted to the hind leg of the stand, and extends forward and alongside of the anvil block. A bar, I, having a series of holes for fastening the lever at any point by a pin. An arm, M, is attached to the lower end of the shaft, C, over the lever, and is connected by a rod, N, to a lever, O, pivoted to the lever F. By moving the lever, O, the shaft is turned and the hammer swung along the face of the anvil.

The hammer handle is connected to a foot lever, Q, by a cord, S ; a coiled spring is fitted, to be contracted when the hammer is forced down, for raising the hammer again. The spring bears against the head of the shaft, C, and the rod connects with the free end of the spring by an adjusting nut. The shaft, C, has a vertical, curved extension which supports a coiled buffer spring that arrests the hammer at the end of the up stroke without shock or jar.

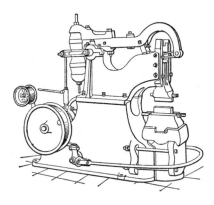

764. BELT-DRIVEN FORGING HAMMER. Bradley type. The hammer in guides is operated by an elastic strap attached to the yoke of a helve vibrated by elastic cushions and an arm with adjustable connecting rod to an eccentric on the belt shaft.

The treadle controls the blow of the hammer by a friction brake.

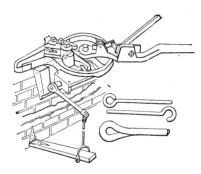

765. EYE-BENDING MACHINE. A hand-operated machine which grips the end of the wire against the central pin by the upper lever, when the lower lever is swung around against the stop and then forming the reverse bend by the treadle and push bar.

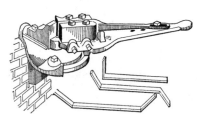

766. ANGLE IRON BENDING MACHINE. The lever, with adjustable jaws and sliding gauge, swings on a pivot and bears against a block. A set piece clamped upon the sector limits the bending angle.

767. PIPE-BENDING MACHINE. The machine consists of a pipe holder, which securely clamps the fixed end of the pipe, and a bend-

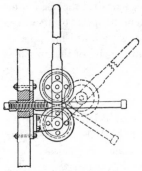

ing lever operated by hand for giving the pipe the proper form. The lever is provided with two grooved wheels pivoted so that the flanges just clear, thus leaving an opening between the bottoms of the grooves of the same size and shape as the pipe. The lower wheel is of such size as to give the pipe the proper radius after it is bent, the bending being accomplished by the upper wheel, which is rolled around the lower one. The device is adapted to bend the pipe without kinking or crushing it, and with one stroke of the lever.

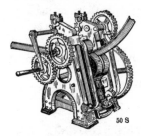

768. ANGLE IRON BENDING MACHINE. For the shape shown the top roller is flat ; the rear under roller is grooved to fit the flange of the angle iron. The front roller is also grooved and is set up for the desired curve by a sliding frame and capstan screw.

769. ROLLED-THREAD-SCREW MACHINE. The cut shows

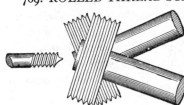

the principle of the rolling process. Screws of $\frac{1}{2}$-inch diameter and less are made with four rollers and are rolled cold. For screws larger than $\frac{1}{2}$ inch three rollers are used and the screws rolled hot.

770. Sample of rolled thread.

771. POWER HACK-SAW. One of the handy specialties of a shop. Automatically cuts off steel bars up to 4-inch diameter. Self-feeding and requires no attention while cutting.

772. SEAMLESS TUBE MACHINE. Mannesmann's process. The principles in this process of making tubing from solid bars of metal

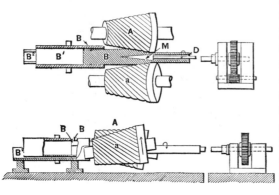

are, that solid bars rolled between a pair of conical fluted rollers, set at an angle, as at A, a, draw the metal from the center of the bar to the outside. The revolving mandril and cone, D, M, smooths the inside of the expanding metal.

B' is the guide tube and frame and B the metal bar.

773. Elevation, showing angle of rollers.

774 and 775. Plan and elevation of gear for rotating the mandril and cone.

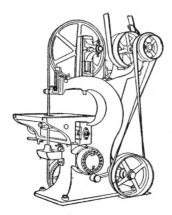

776. METAL BAND-SAW. The modern metal band-saws are made to cut solid bars, such as center cranks, Y's in connecting rods and locomotive frames, as well as structural forms of all kinds.

777. HAND-SCREW TIRE-SETTING MACHINE. The blocks surrounding the tire are set up by hand screws and a wrench, compressing the tire tightly upon the wheel.

778. HYDRAULIC TIRE-SETTING MACHINE. A number of cylinders and pistons set within a strong iron ring compress the tire

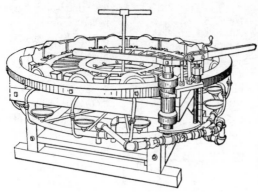

upon the wheel by the power of the two pumps connected to the cylinders by pipes. The large pump is for filling the cylinders and bringing them to a bearing upon the tire, after which the small pump is operated for great pressure.

779. AUTOMATIC FURNACE for hardening and tempering balls. The balls are picked up by the small shelf in the revolving hop-

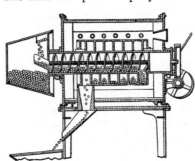

per, and tipped into the mouthpiece, and carried along by the central screw; transferred into the outer and hotter reverse screw carrier to the chute, and dropped into a water bath. The chamber is heated by gas jets, which can be regulated for temperatures suitable for hardening or for drawing temper.

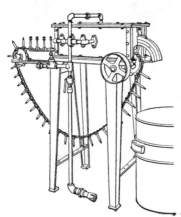

780. GAS-HEATED HARDENING AND TEMPERING FURNACE for small articles, as bicycle cones, shells, or any articles that can be placed on the pins and carried through the furnace at the proper speed for the required temperature and dropped into the water bath. The heat is regulated by the gas and air valves.

American Gas Furnace Co. type.

781. TEMPERING BATH. A pot of oil or tallow is set in a gas-fired furnace inclosed so that the flame can not set fire to the oil vapor. A thermometer immersed at the side shows the proper temperature for the desired degree of temper.

782. DOWN-DRAUGHT GAS-MELTING FURNACE. The burner B, of combined gas and air, enters the flame at the top of the crucible and discharges to the chimney below the bottom of the crucible at H. E is the cover lifted by the lever C, and chains to swing off the furnace. The hearth is a perforated fire tile on which the crucible sets.

American Gas Furnace Co. type.

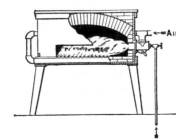

783. OIL OR GAS FIRED FORGE. Oil or gas enters the atomizer by the small pipe and is mixed at the nozzle by a strong blast of air. Additional air jets to complete the combustion enter beneath the bed of the furnace.

784. MELTING FURNACE for brass, copper, or bronze. Operated by gas or crude oil, and compressed air.

Oil or gas is fed through the small pipe, atomized and mixed with air in the inlet nozzles at the top of the cupola and the flame projected down upon the metal. The cupola is tipped by the wheel and gear to pour the metal from the side spout.

785. DUPLEX MELTING FURNACE. Rockwell type. No crucibles are used, the furnace chambers being lined with refractory

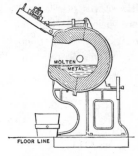

material which is inexpensive and cheaply applied, and the charges of metal to be melted being placed in the chamber, as in the right-hand chamber, 786. The fuel used is oil or gas, the air being supplied by an ordinary fan or pressure blower and there being a burner at each outer trunnion. But one of these burners is normally in operation at a time, the flame which is melting one charge extending into the other chamber and giving up much of its remaining heat to the fresher charge of metal. When the charge in either

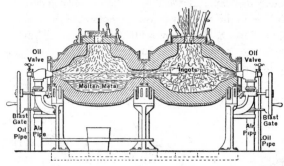

chamber is completely melted it is passed out by turning the chamber and bringing the mouth down to the pouring position. The two chambers may be used for different metals or for the same metal, and both charges may be melted so as to be poured together if a large quantity of metal is required at once. The halves of the chambers are hinged so as to make the entire interior perfectly accessible for relining or for any purpose.

786. Longitudinal section of the double furnace.

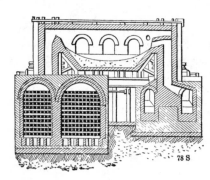

787. OPEN HEARTH STEEL FURNACE, showing the concave hearth working doors and the regenerator ovens, which heat the incoming air that feeds the furnace.

788. HOT - METAL · MIXER . Rolling type. Designed for a capacity of 250 tons. The vessel is composed of steel plates formed in

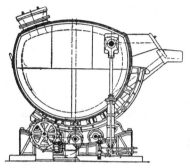

cylindrical and spherical segments, and requires no additional bracing. It is lined with best magnesia bricks. The vessel rests on two circular roller beds, each composed of five rollers, supported by pedestal bearings, resting on foundation girders. The roller tracks fastened to the vessel are of cast steel. Concentric with these are two rack segments of cast steel, by which the mixer is tilted. The pinions meshing into the segments are driven through gear trains from a 26 horse-power electric motor. An additional tilting device is provided, consisting of a vertical hydraulic cylinder at either side, with their plungers linked to pins projecting from the sides of the mixer vessel near its front end. A precaution is provided in the form of a hook attached to the rear face of the vessel, to which the regular traveling crane serving the mixer may be hitched and the vessel thus tilted.

789. HOT-METAL MIXER. Tilting type. The hot-metal mixer shown in the cut is designed for a capacity of 275 tons of fluid metal.

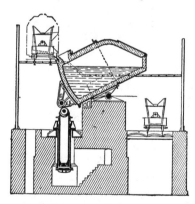

The maximum external dimensions of the containing tank are about 15 feet diameter and 27 feet length. The shape is cylindrical, with a conical pouring spout at the front end, which converges from the full width of the tank to a narrow opening. A charging funnel is on the top of the tank at the back end. The tank is lined with magnesia bricks to well above the slag level. The support of the tank is a large pin, resting between two saddle castings bolted respectively to the tank body and the foundation. A cast chair built into the foundation forms a rest for the heel of the tank when this is tilted back. The molten metal is charged into the filling spout from ladle cars running on an elevated track back of the mixer. The metal is run from the mixer into other ladle cars running on a platform at a lower level, which extends entirely around the mixer. These ladles then run directly to the converters or the furnaces.

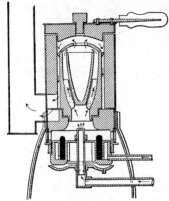

790. KEROSENE-OIL MELTING FURNACE. The small pipe supplies oil to an annular wick of asbestos. The combustion chamber has air holes around the outside with dampers to regulate the air supply. The central tube is the compressed air supply from a blower to give force to the flame to drive it around the crucible and down the annular chamber to the chimney.

791. PETROLEUM FORGE for heating rivets. The rivets are introduced through the door *a ; b* is a movable cover, which is dis-

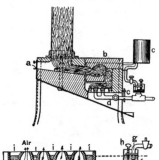

placed in order to remove them from the forge ; *c* is the device that supports the burner *d*. This latter consists of a row of receptacles, *i, i,* in which the liquid fuel is kept at a constant level through a small reservoir, *j,* which receives the inlet tube, *g,* fixed to the closed reservoir *c*. A small screw, *h,* permits of regulating the depth of the oil in the constant level reservoir, *j,* and burners by raising or lowering the mouth of the tube *g*.

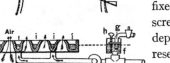

792. Section showing regulating reservoir and burner cups.

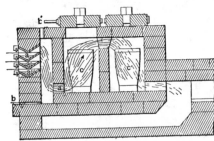

793. PETROLEUM MELTING FURNACE. Nobel type. *a, a', a''*, oil-burner troughs. *b, b'*, air regulating inlets. *c, c'*, crucibles.

Fire flue at bottom. See Fig. 147 for details of the burner.

794. PETROLEUM FIRED REVERBERATORY FURNACE.

The petroleum enters the troughs, *a*, of the reverberatory furnace through

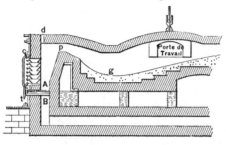

the pipe *c ; t* is the pipe through which passes the overflow of the basins, and *d* is the air port designed to regulate the combustion. The flame breaks against the fire bridge, *p*, before reaching the furnace bottom, which is composed of quartzy sand and clay. In case of stoppage of the work, the flame proceeds toward the flue, B, which is normally covered by a stone A. The casting is effected through the tap hole *g*.

The charge is put in, as usual, at the back of the furnace, and is made to advance progressively through the working holes.

See Fig. 147 for details of the burner.

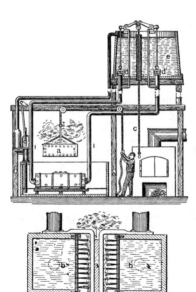

795. PLATE HARDENING MACHINE. Urban's type. For uniformly hardening steel plates and armor plates without risk of bending or buckling.

For this purpose, the heated plate, *a*, is lowered until it rests vertically between guides midway between two tanks, *b*, *b'*, having numerous perforations in their sides next to the plate. By pulling a cord, *c*, the valves, *d*, *d'*, are opened, and salted water from the reservoir, *e*, descends through the pipes, *i*, *i*, into the tanks, playing in jets against-both sides of the plate ; a pump returns the water from the lower reservoir to the upper one, in order that it may be used again.

796. Section showing the plate held in place by guards and the water jets playing upon it.

797. DOVETAILING MACHINE. Plan and elevation of a machine in which the work is done by a gang of saws on a mandrel. The

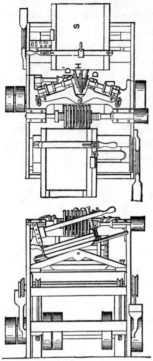

mortise-cutting portion is the right-hand part of the lower figure. In it the board is secured on the carriage, S, in such position that the edge of said board projects under the saws or cutters more or less, according to the depth that the dovetailing is to be cut, which will be governed by the thickness of the stuff. The board, on being properly adjusted, is then brought in contact with the saws by elevating the table, thereby carrying the board upward to the saws, D, D', cutting the sides of the mortise, and of any angle that may be required, by adjusting the stays in which the cutters are hung to the required angle.

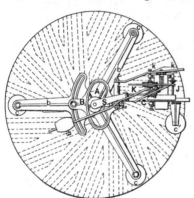

The central cutter, H, as will be seen, cuts into the board at a right line between the side saws, and as it leads in the cutting, the central portion of the mortise is cut away ; the side saws, as they follow, cut away the remainder, leaving a clean, angular mortise for the admission of the tenon.

798. Elevation, showing saws and angle of board to be dovetailed.

799. DIAMOND MILLSTONE-DRESSING MACHINE. A Swiss machine for dressing millstones. The frame, A, has arms, *b*, *b*, terminating in feet, *c'*, which are provided with set screws. A tool support, S, is pivoted to the center of A, and is adjustable by means of sector, B, and slides on the arm, C, of the frame. Two disks at K carry diamonds on their peripheries, and are set in rapid revolution by belts from spindle, J, which is revolved from any convenient shaft outside the millstone.

The cutting disks being put in

rapid revolution, the successive blows of the diamonds act in a manner similar to that of a hand tool, and parallel grooves are cut in the face of the stone. Three of these sets of parallel channels or grooves make one division of the stone. The guide bar, C, is adjustable, so that the stone may have a right-hand or left-hand dress, as desired.

800. FILE-CUTTING MACHINE. The slide on which the file is bedded oscillates laterally so as to adapt itself to the variations of the

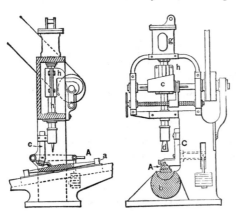

surface of the file. For this purpose the slide, a, is made convexly cylindrical at the under side, and is supported in a concave guide, b, in the bed frame of the machine. The file, A, is caused to present its surface parallel to the cutting edge of the chisel, d, by a plunger, c, sliding freely in a guide, the plunger, c, carrying at its lower extremity a feeler blade whose edge rests upon the surface of the file. Contact between the blade, c, and file, A, is insured by means of a weight acting upon the plunger c. The feeler blade is thus free to accommodate itself to variations in the surface configuration of the file, but being held rigidly in the transverse direction, compels the file to accommodate itself laterally to the blade and so presents its surface on the line of cut, truly parallel to the cutting edge of the chisel d. The chisel, d, delivers its blow under the impulse of a spring in a casing, g, the spring being compressed at each stroke by the upward movement of the ram which is alternately lifted and let fall by a revolving cam engaged by an arm h. Variation in the degree of compression of the spring, so as to produce any desired graduation in the strength of blow of the chisel, d, is brought about by increasing or diminishing the effective radius of the cam to increase or diminution in the height of lift of the ram h. The cam is made tapering in the direction of its axis and is mounted upon its shaft with a groove and feather connection so as to be longitudinally adjustable in order to bring any portion of its length to act upon the arm h.

801. Front elevation showing details of the machine.

802. DOVETAILS. The three upper figures show the method of end splicing by dovetails.

The series of illustrations show the several modes of dovetailing the edges of boxes and drawers.

o is a *miter* and *key* joint.

p, the common dovetail joint.

q, the *half-lap* dovetail.

r, the *secret* dovetail.

s, the *lap* dovetail.

t, the *miter* dovetail.

a shows the ordinary dovetail with the parts detached ; *b* the parts put together.

Concealed dovetails are made in two ways :

c, *d* show the *lap dovetail*, in which a fin of wood on the return edge hides the ends of the tenons and mortises.

803. MORTISING DOVETAIL MACHINE. The upper bed surface consists of two equally but oppositely inclined planes, B', B",

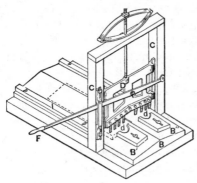

whose slope corresponds with the chamfer of the desired dovetails. C, C are standards guiding in a vertical path gate D, in which is fixed a series of chisels whose cutting ends are at such an unequal elevation as to correspond with the obliquity of the planes, B', B". These chisels are readily adjusted to any height and degree of separation, and are fixed to their proper positions by screw bolts.

The gate is elevated and depressed by means of a lever, F, and is gauged or arrested in its descent by a stop or shoulder. Stops on the planes, B', B", gauge the stuff. I is a gauge for the edge of the stuff.

The board containing the heading pins already sawed is placed on one of the inclines, B', B", and the chisels, being caused to descend, operate to excavate on one side the intervening stuff between the pins. The stuff being then placed on the other incline, and the gate again depressed, the excavation is completed by cutting away the opposite sides.

For excavating the mortises, the doubly inclined block, B, is removed, and another gate substituted for the gate, D, in which substitute gate the chisels are so secured as to have their lower ends in a horizontal line. The stuff being placed on the horizontal bed and the chisels depressed, the surplus timber is excavated at a single stroke.

804. FILE-CUTTING MACHINE. The sliding head to which the shank of the blank is clamped is actuated by a feed screw and half

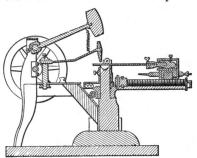

nut, the latter being automatically raised to stop the feed motion at the proper time. The anvil has a hemispherical block, whose convex side rests in a socket of its support. The anvil and feed movement are supported on a turntable, by whose adjustment the inclination of the teeth is determined. The chisel is supported upon a flexible rod, which is connected to the hammer handle by a spiral spring. The hammer is attached to a rock shaft, which has an adjustable arm acted on by a cam on the main shaft.

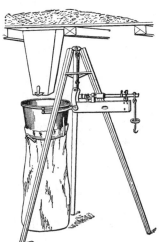

805. BAGGING AND WEIGHING SCALES. A tripod on which is fixed a Roman balance with an extension yoke and funnel to which the bag is attached by a band and clips. The beam of the balance is made to counterbalance the yoke and funnel, and the tare of the bag is placed on the hook at the end of the scale beam.

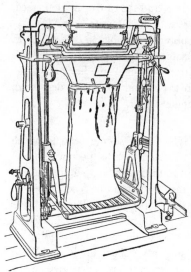

806. AUTOMATIC BAGGING AND WEIGHING MACHINE. The bag is attached to the hopper funnel with its bottom resting on the grated weighing platform. The feeding valve is then opened and connected by an automatic latch to the weighing platform, which drops at a set weight of filling, disengaging the latch, when the valve closes on the feed spout. The adjustment is somewhat complicated and allows for the average tare of the bags.

807. TURPENTINE STILL with by-products of creosote, pyroligneous acid, etc., produced by the destructive distillation of wood.

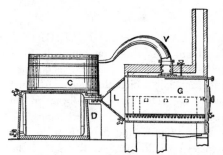

The blocks of wood are placed in the retort, G, the door on the right closed, the valve at the conical end closed, and the valve above opened. Water being introduced into the chambers to the level of the grate bars, fire is applied, and the clear white spirit passes in vapor by the neck, V, to the worm in the tub C. As soon as it begins to show color, the valve above is closed, and the valve at the conical end opened, when the vapor passes through the purifier, L, into the chamber, which is surrounded by water in the tub D. In this manner the different distillates are kept separate in the several receivers.

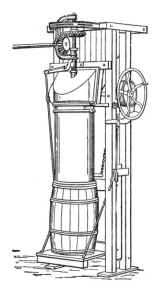

808. FLOUR PACKER. A chute with a quick-closing gate delivers the flour to the barrel, in which a revolving auger propeller packs the flour tightly. A sleeve guides the flour and prevents waste.

The packing propeller is at the bottom of the sleeve, and the filling commences with the barrel elevated by the movable platform and hand wheel, so that the propeller continually acts upon the surface of the compressed flour while the barrel gradually descends to the floor.

808A. FREIGHT CARRIER. A device for the rapid handling of freight is shown. The machine consists of an endless chain of

special construction provided with pivoted hooks, which are carried by small iron wheels, which roll upon the flanges of a pair of channel beams bolted together and spaced about 1 inch apart, thus forming a strong truss frame or carrying track capable of spans of considerable length. At each end the chain passes over a sprocket wheel, and these wheels can be driven by motors in either direction. The method of loading is illustrated very clearly.

SECTION XIX.

TEXTILE AND MANUFAC-TURING DEVICES, ETC.

Section XIX.

TEXTILE AND MANUFACTURING DEVICES, ETC.

---

809. PATTERN BURRING MACHINE for figured woolen goods. German design. *a* is a revolving metal brush, under which the stencil

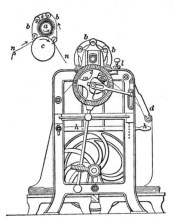

plate, *b*, passes as an endless sheet, guided by the small rollers *f, f, g, g.* The cloth passes underneath the plate, and has the same speed as this plate; it is carried forward by the guide-roller, *c*, which at the same time presses it against the plate. The driving pulley is on the main shaft, from whence the motion is passed on through a diagonal shaft and bevel wheels to the roller, *c*, on one side, and by a belt to the brush on the other side. The brush and the guide roller, *c*, run in opposite directions. The brush is covered in its upper half by a cast-iron cover, which protects it against injury, and at the same time keeps it in the bearings. These bearings are made to slide up and down, and are pressed upward by a set-screw, *l*, acting upon one arm of a lever whose other arm has a pin pressing on the under side of the bearing. A batching apparatus driven by a heart cam motion completes the machine.

The operation of the machine is as follows: The cloth, which has been milled and raised in the usual manner, is introduced with one end between the stencil plate and the guide roller, *c*, and with the nap running in the same direction. The brush, which must be set so that the wires project through the open places in the plate, and ought to run at a high velocity, raises the nap on these open places in a contrary direction to that of the existing nap, and this forms the design, which may be either the roughened or smooth surface.

810. COTTON-SEED HULLING MACHINE. A machine by which the hull of the cotton-seed is rasped off by the two corrugated wheels and sifted by the revolving shaft and screen from the fari-naceous and oily matters, which are utilized for their oil and the refuse for manure.   The kernels pass through the screen, while the coarser hulls and fibers are carried along and discharged from the lip of the screen.   The hulled seed is then re-

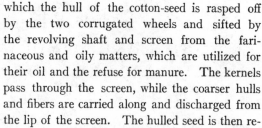

ceived into the box-screen I, which, being shaken by suitable mechan-ism, separates the still remaining ·lighter portions of the hulls that pass the wire screen, carrying these portions out over the apron J, while the cleaned and hulled seed passes out through the chute K.

811. COTTON BAT COMPRESSOR AND CONDENSER. *a*, lint flue ; *b*, condenser drum ; *d, d*, bat-former aprons ; *e*, compression roll ; *f, f*, baling rolls ; *g*, core ; *h*, baling belt ; *i*, belt idler ; *j*, hydraulic cylinder ; *k*, pres-sure column ; *l*, press pulley ; *m*, bat-former pulley ; *n*, pis-ton rod ; *p, r*, tension rolls ; *s*, pressure gauge ; *w*, guides for idler ; *x*, bed plate ; N, pressure regulator.   The con-densing drum spreads the lint evenly ; t h e aprons press it between the bat-f o r m e r pulleys, then passing un-der the com-pression roll,

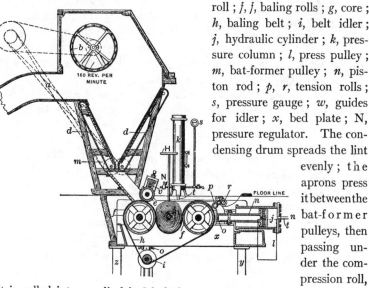

it is rolled into a cylindrical bale between the baling rollers.

812. COCOANUT-PARING MACHINE. A train of gearing is arranged to revolve the nut and the circular plate carrying the par-

ing device. The circular plate is secured to a sleeve on the central post projecting from the base, and receives its rotary movement from a bevel gear mounted on the sleeve. On the post immediately above the sleeve carrying the circular plate is a collar carrying the horizontal arm which supports the knife post, and works over the face of the circular plate. The horizontal arm has a depending lug at its outer end which engages in turn with opposite marginal apertures in the circular plate. The post carrying the box in which the paring knife is held is jointed to the horizontal arm, and at the junction of the two is a coiled spring to force the post against the nut. A coiled spring is also placed on the central post, one end being secured thereto, and the other end to the horizontal arm. In operation, the lug in the horizontal arm being in the aperture on the right, the circular plate carries it in revolving and also the knife post ; as the knife reaches the end of the paring on the left, the arm rides up on a beveled lug on the case, which forces the lug on the arm out of the aperture, and the coiled spring on the central post retracts the arm back to the first position at the right.

813. FLOCK GRINDING MACHINE. In this machine the feed box has radial agitators on a vertical, rotating shaft. The endless

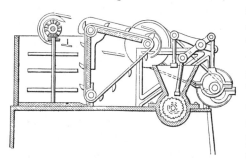

apron passes up one side, and has cups carrying up the material and conveying it to the hopper of the tearing cylinder. The material is forced down upon the tearing cylinder by reciprocal plungers and, carried along the fluted cutter, is discharged into a box at the side of the machine.

814. FLAX-SCUTCHING MACHINE. For threshing and scutching flax.   The stalks are fed from the table B between two fluted

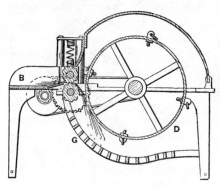

rollers, the lower one of which is journaled in fixed bearings, and the upper is yielding, being pressed down by spiral springs.   On passing through the rollers the stalks are subjected to the action of a series of swinging beaters pivoted in eye bolts on the drum D, which rotates at about ten times the velocity of the rollers.   The separated seeds drop through the slatted bottom G, and the bruised fiber is conveyed to an opening at the rear of the machine.

815. MULTIPLE-STRAND CORDAGE MACHINE.   The eighteen bobbins with which it is provided are each armed with a special

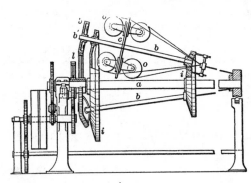

brake which can be regulated with the greatest precision, so that, during the reeling, the tension of the yarn remains invariable, this being an important point.

These bobbins are mounted upon three disks (three in front of and three behind each disk), which revolve between two others keyed upon the central axis.   When the machine is in motion, the three disks are carried along in a certain direction ; but, by the combination of the gearings, they revolve at the same time around their axis in an opposite direction.   Each strand therefore receives the same tension.

When the bobbin yarns have each been twisted upon its own laying-top, the three principal strands that they have formed pass to a central laying-top, where they are twisted together.   Thence they are carried along by polished friction drums and wound upon reels.

816. PAPER ENAMELING MACHINE.   For glossing paper and card stock.   The enameling mixture is thoroughly stirred within

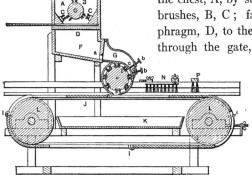

the chest, A, by stationary and revolving brushes, B, C; falling through the diaphragm, D, to the chamber, F, it passes through the gate, *a*, to the spout, G, whence it is admitted by the adjustable screw valves, *b*, *c*, to the brush roller, I, rotating within a cylinder having an opening at its lower side, from which the brushes, N, spread the enamel on the sheet of paper passing along on the endless belt J.   This is revolved by two cylinders, L, L, having fingers, *i, i*, which clutch the sheet when presented to them, and after carrying it past the spreaders, N, and blenders, P, fall and release it.

817. CORDAGE-MAKING MACHINE.   Modern type.   A three strand, multiple thread machine in which the cable spool is revolved

for twisting the three strands issuing from the triangular eye frame, all the moving parts being automatic and driven from the pulley at the rear of the spool head.   An illustrated description of the details of this intricate mechanism is not available; but the subject is a valuable study.

818. THREE-STRAND CORDAGE MACHINE. The armed carrier wheels, K, H, T, are fast on the driving shaft, S, which is driven

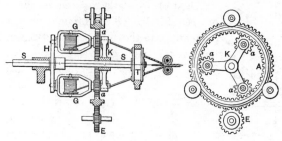

by a gear wheel meshed with the three pinions revolving the spools, G, for a back twist of the strands. The pinion, E, revolves the ring gear, A, in guide rolls, the twist of the strands and cordage being in opposite directions.

819. Cross section, showing back twist spool gears and arm, K.

820. THIRTY - TWO STRAND CORDAGE MACHINE. A fine study of the mechanical motions required in this complicated

mechanism for the manufacture of so simple a thing as a rope. Four strands, each composed of eight yarn strands.

The tension of the finished rope and its winding on by the reel at a uniform rate with the twisting speed of the machine is a most important feature ; and is operated by means of two grooved drums placed tandem, over which the rope is wound twice to give it a frictional pull. The drums are rotated by a fore and aft shaft and gearing from the main driving shaft. The reel is revolved by a friction belt, which allows for varying its speed for equal tension of the rope as the reel fills up.

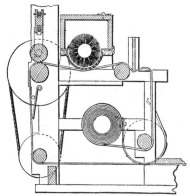

821. FLOCKING MACHINE. For distributing flock evenly on a prepared surface of cloth or paper.

The cloth or paper is passed on an endless web with its varnished or glued side uppermost, the varnish or glue being applied by an elastic roller fed from a hopper not shown. The flock is evenly fed to the surface of the cloth or paper by the revolving brush in the hopper at the top of the machine.

822. ELECTRIC CLOTH CUTTER. A revolving sharp-edged blade driven by a motor in the head frame on an arm or tripod. It is

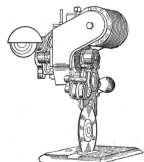

14 inches in height and weighs 35 pounds. It is capable of cutting any thickness of cloth up to $3\frac{1}{2}$ inches and any width or length. A feature of the cutter is the fact that it is perfectly portable, so that goods on any of the cutting-room tables can be cut with it. This is a valuable feature, as it obviates the folding of the goods and the carrying to the machine.

In order to keep a perfect cutting edge, grinders are attached and can be brought into contact with the knife in an instant. Wolf Electric Promoting Co., Cincinnati, Ohio.

823. QUARTER SAWING OF LUMBER. Three methods of sawing lumber, one of which, at the left, is the ordinary method, the two

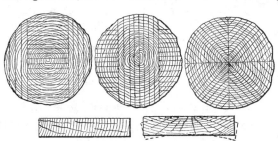

at the right representing the cuts for quarter sawing. An example of a piece of quarter sawed and common sawed lumber is shown in the lower figures. The dotted lines in the common sawed piece indicate the curl in drying.

824. EVOLUTION OF THE LAG SCREW and the machine for making them, showing the general construction of the cutter and

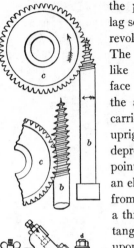

the principle upon which it operates, b being a lag screw which is being cut and c the cutter, each revolving in the direction indicated by its arrow. The lag screw and the cutter run together, just like a worm and a worm wheel. The cutting face of the cutter is parallel to and in line with the axis of the lag screw. The cutter spindle carrier, k, is carried by horizontal trunnions in uprights on the carriage. It is obvious that a depression of the arm, k', will bring the cutting points in toward the center of the lag screw, while an elevation of the arm will swing the cutter out from the center of the lag screw, or cause it to cut a thread of larger diameter. The bar, n, of rectangular section, is fastened rigidly to an upright upon the frame, so that when the carriage moves along it slides over this bar. The under edge of the bar is not straight, and against this edge works the roller, m, in a fork rigidly attached to the arm k'. Soon after the cutting of the screw begins, the roller, m, comes to a portion of the under edge of n, which curves upward, and this upward curvature, of course, allows the arm, k',

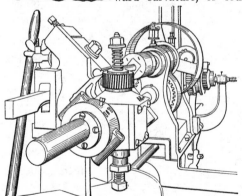

to rise and the cutter to swing away from the center of the lag screw, thus forming the taper point. A straight portion of n then forms the parallel portion of the lag screw, and a further rise allows the cutter to clear the screw entirely, when it drops out finished, and the carriage runs back, bringing the cutter in position to begin another cut. Roller m is held in contact with bar n by the pull of spring o.

825. Details of cutter and guide block.

826. General view of the screw cutting machine.

827. PORCELAIN MOLDING MACHINE. French model. The apparatus, of which a front view is given, consists of a vertical frame

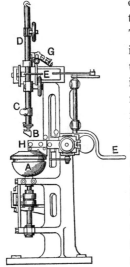

carrying the lathe below, a calibrating tool in the center, and the molding tool above. The chuck, A, coming from a second machine is secured to the lathe head. B is the molding tool moved by the handle C. D is an adjusting collar. E a carriage regulating the movement of the tool effected by the handle E. G is a gauge for regulating the form of the plate. H is the calibrating tool. The chuck being on the lathe head the tool is caused to descend, and this meets the paste at the center, determining its thickness. Being restricted in its motion by the guide or gauge G which represents the profile of the plate, and being submitted to a horizontal movement, it necessarily works the object according to the desired exterior form indicated by the gauge.

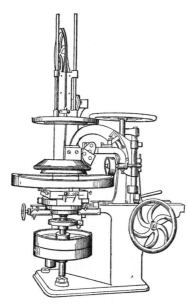

828. PORCELAIN MOLDING MACHINE. French design. The pulleys on the vertical spindle operate an oval chuck, while a curved form is given to the molded disk by the revolution of the curved ring and roller which moves the trowel for the required shape of the disk. The hand wheel operates the various movable parts and the trowels for molding the shape to the form of the curved disk.

829. DIAMOND CUTTING.   The form a diamond shall assume is determined by its shape in the rough, the duty of the lapidary being

to cut it so as to sacrifice as little as possible of the stone and obtain the greatest surface, refraction, and general beauty. Having decided upon the form, a model is made in lead and kept before the workman as a copy. The rough diamond is cemented with fusible metal to a handle called a dop, *a*, leaving the part exposed which is to be removed to form one facet. The projecting portion is then removed by attrition against another diamond similarly set in a handle, B, and finished by means of diamond dust and oil upon a steel disk or wheel, according to circumstances.   When a facet is finished, the stone is reset in the handle and the process repeated. Several months are expended in cutting large stones, as the work proceeds very slowly.

The polishing is performed upon a rapidly revolving steel wheel, *d*, driven by a band, *g*, and fed by hand with diamond dust and oil.   The diamond is set in a dop as before, on the end of a weighted arm, *f, e*, and held against the wheel, the results of the process being collected in a box for future operations.

Diamonds with flaws or imperfections are sawed asunder or split, the latter (shown at A) being a speedy but risky operation, requiring great judgment in determining the plane of cleavage and skill in the use of the chisel, *b*, and hammer.   For sawing, a fine wire is used, fed, as in the case of the revolving wheel, with diamond dust and oil.

830. Angle gauge for observing the angle of the facets.

831. DIAMOND CRUSHER AND MORTAR. Diamonds for the use of the lapidary are crushed in a mortar, which consists of a

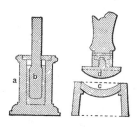

cylindrical box, *a*, and a pestle, *b*, both made of hardened steel. A small rough diamond is placed in the mortar, and the pestle driven down by a hammer. The pieces of broken diamond are examined for the detection of fragments suitable for gravers, drills, and etching points. The remainder is crushed to an impalpable powder by several hours' continued work, rotating the pestle between blows.

When sufficient fineness is not attained by the mortar, the dust may be ground between the concave and convex surfaces, *c*, *d*, of a hardened steel mill, a little oil being added to the dust. The particles will grind each other.

832. Section of the grinding mill.

833. DIAMOND HAND TOOLS AND DRILLS. In Fig. *a*, *a* are front and side views of diamond chisels used in turning rubies for

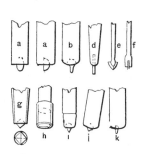

watch-jeweling ; *b* is a diamond drill for making the hole in the ruby plate ; *d* is a tool of steel wire to be used with diamond dust in drilling jewels ; *e*, *f* are two views of a triangular fragment of diamond mounted for drilling china or porcelain ; *g* is a square stone mounted for the same purpose ; *h* is a metallic tube for drilling annular holes in jewels with diamond dust ; *i* is a diamond point mounted for etching or ruling in engraving ; *j*, *k* are diamonds mounted for ruling graduations of mathematical instruments.

834. COMBINATION PRESS for fruit, lard, or a sausage stuffer. For fruit, the gauze wire basket and strainer diaphragms are used. The piston, screw spindle, and gear are attached to the swiveling yoke and leave the cylinder clear for charging and cleaning.

835. ARTIFICIAL FLOWER-BRANCHING MACHINE.

French type.   The basis of the stems is wire, and two threads of suitable

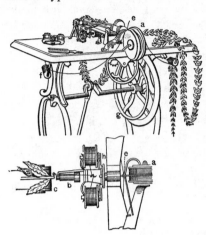

material are laid along this wire to prevent subsequent slipping of the colored thread which forms the outer covering of the stems. The ends of the short stems of leaves, flowers, buds, and fruit being laid against the wire are wound under the outer covering, and are thus fastened to it.

The wire is fed from a spool, *a*, passes through a hollow spindle, *b*, and lies upon an endless feed belt, *c*, to which it is clamped by small pinchers.

The belt is driven by gearing underneath, and carries with it the wire stem, which is slowly unwound from the spool *a*.   Two threads, passing through an eye, *e*, are also drawn through the hollow spindle, *b*, in conjunction with the wire, by the motion of the endless belt.   These threads are unwound from the spools *f*.   At the same time a rapid rotary motion is given to the hollow spindle by a small belt from the driving pulley *g*.

On the revolving hollow spindle, *b*, is fixed a spool frame, *h*, which carries two spools.   The covering threads are led from these spools through the loop of a small flyer on the end of the hollow spindle, *b*, and being held in contact with the wire as the latter is slowly fed through the spindle, are wound uniformly over its surface, the spool frames revolving with the spindles.

The ends of the stems of leaves, fruits, or flowers being thrust into the ends of the hollow spindle are at once caught, and firmly wound under in a rapid manner.

836.   Detailed figure of the winding operation.

SECTION XX.

ENGINEERING AND CON-STRUCTION, ETC.

Section XX.

ENGINEERING AND CONSTRUCTION, ETC.

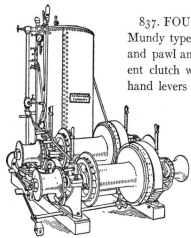

837. FOUR-SPOOL HOISTING ENGINE. Mundy type. Friction drums with stop ratchet and pawl and friction-brake straps. Independent clutch winches. All under control of four hand levers and two foot levers.

A most convenient type of combination hoist where a great variety of work is in progress. The two drums and four spools have each an independent motion and stop.

838. DISINTEGRATOR. Blanchard type. On two concentric shafts driven at high speeds are mounted grids or cages, one within the

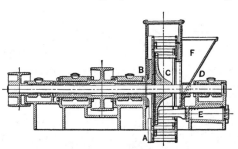

other, driven in opposite directions. F, the feed hopper. E, a steel pin projecting within the inner cage to receive the impact of the coarse material. A, the outer cage disk frame. C, inner cage disk frame. B, casing. D, D, B, journal boxes. Velocity of the periphery of the cages about 6,000 feet per minute.

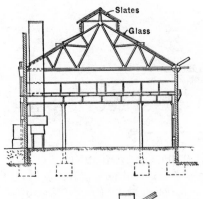

839. FOUNDRY CON-STRUCTION. Steel construction with cupola in the main room. A platform or floor above and covering part of the molding floor in a confined building, is made available for fuel and iron storage and for feeding the cupola. A molding floor with cupola in a steel frame extension arranged to draw the metal on the molding floor.

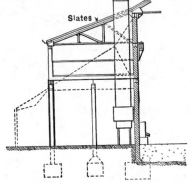

840. The extension may be also a power house with boiler, engine, blower, hoist to the charging floor and storage for material.

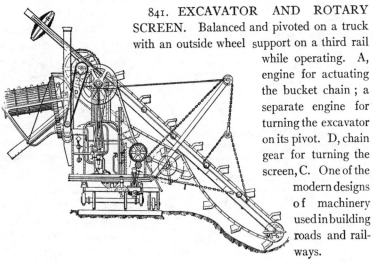

841. EXCAVATOR AND ROTARY SCREEN. Balanced and pivoted on a truck with an outside wheel support on a third rail while operating. A, engine for actuating the bucket chain ; a separate engine for turning the excavator on its pivot. D, chain gear for turning the screen, C. One of the modern designs of machinery used in building roads and rail-ways.

842. UNIVERSAL POCKET LEVEL. The under side of the glass is ground and polished spherically, concave of long radius, and set in a case of steel, nickel-plated. Filled with spirits or glycerized water, except the bubble space.

843. ADJUSTABLE B E A M C L A M P. For suspending iron pipes from fireproof ceilings. The sleeve when turned into a quarter

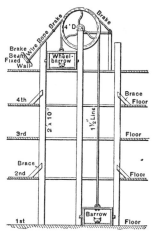

turn allows the clamp to be adjusted, and is then locked by turning it back to its original position, the teeth on the hooks engaging in corresponding slots on inside of sleeve. The hooks are sharpened where they extend over the flange· of the beam so they can be driven under the brick.

844. GRAVITY ELEVATOR. A simple arrangement for lowering building material in taking down high buildings. When the upper story is cleared the wheel is set on the next floor below and so on. The brake controls the difference in weight between the empty and loaded barrow.

This method of lowering the material in taking down the old buildings, avoids the dust nuisance made by discharging the material through a chute.

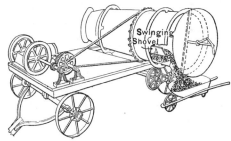

845. PORTABLE CONCRETE MIXER. A rotary mixer driven by a steam or compressed-air engine, with swinging shovels on the inside of the barrel for thoroughly mixing the concrete.

846. CONCRETE MIXER. Smith type. Mounted on a truck and driven by a gasoline engine. Mixes in batches and tilts to discharge while running.

847. PORTABLE CONCRETE MIXER. Square box type. The

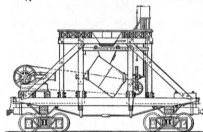

revolution of the rectangular box, hung at its corners, makes a thorough mixture of the concrete in batches. The materials are charged through the hopper in measured quantities, so making a uniform mixture for concrete work.

848. TRENCH BRACE. An up-to-date contractor's appliance for

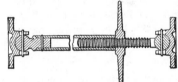

bracing trenches. The large handle nut and screw give the brace great power, and the socket bearings accommodate the brace to irregular surfaces.

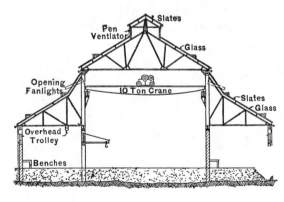

849. TYPES OF MACHINE-SHOP CONSTRUCTION. Sides may be of brick or steel with corrugated iron siding. Roofs of steel framing with slate covering and glass lights in roof.

850. WOOD PRESERVATION APPARATUS. Hot-air and tar-vapor process. Heated air is driven from the generator, D, into the

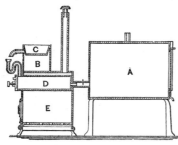

chamber, A, containing the wood, the vapor escaping from the upper pipe. When the wood is dry, tar is introduced into the generator, and the resulting fumes similarly forced into the chamber impregnate the wood.

B is a water box made hot from the furnace, E, which in turn keeps the tar fluid in the tank C.

851. WIRE-GUY GRIPPER. The eccentric grooved levers, as shown, make a quick-handled grip on guys for derricks. It is easily applied or removed by the use of the pin in one of the lever sheaves.

With the addition of parallel jaws under the eccentric grips, this device makes a good grip for hauling ropes and cables.

852. TIMBER CREOSOTING APPARATUS. Timber or piles are bundled and shoved into a long cylinder and the cylinder head

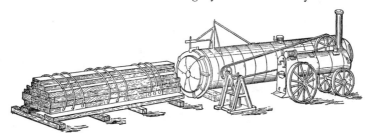

closed tight. Steam is then introduced at a high **pressure**—100 to 150 lbs. per square inch. This heat coagulates the sap and drives the moisture from the lumber, when creosote oil is pumped into the cylinder and saturates the wood. The oil is then driven out of the cylinder by the steam pressure and heat and the lumber withdrawn.

853. ELECTRICALLY DRIVEN HAMMER. Power is transmitted to the crank shaft, A, by means of a flexible shaft, and a recipro-

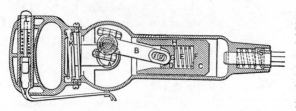

cating motion is given to the hammer head, C, by the pitman B. As the rotation of shaft A is very rapid the succession of blows upon the chisel is so rapid as to be almost continuous. A small balance wheel on the end of shaft A absorbs part of the shock of the impact and produces smooth running. A plunger, D, is free to slide within the hammer head, but is kept from striking point F by the coil spring at E. When the pitman moves to the right its right-hand end presses against the plunger at G, which in turn communicates motion to the hammer head through spring E, and the chisel is struck a sharp blow.

After the blow, the hammer head is returned by the movement of the pitman to the left, by means of pin g. This pin is attached to hammer-head C, but is entirely independent of the plunger D. The end of the pitman is slotted to receive pin g, and the slot is of such a length that the pin, together with the hammer head, can have free motion during the blow.

854. DUPLEX ROLLING LIFT BRIDGE. Scherzer type. These bridges cover a double waterway of 110 feet each, over Newark Bay, on the line of the Central Railroad of New Jersey.

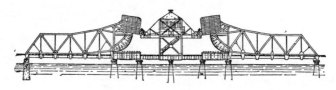

Two 75-horse-power gasoline engines, manufactured by Fairbanks, Morse & Co., are provided. Each engine is so arranged and connected with the machinery that it can operate both bridges either jointly or separately, as desired. The operator's house is constructed entirely of steel and fireproof material, wood being eliminated.

855. BALANCED SWING BRIDGE., Toronto, Can. Operated at the short end by an endless chain guided by pulleys on a quadrant

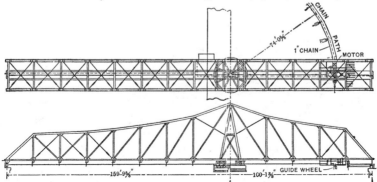

frame. The bridge is 160 feet on the long span and 100 feet on the short span. Balanced on friction rollers by counterweights and operated by electric motor. Plan and elevation.

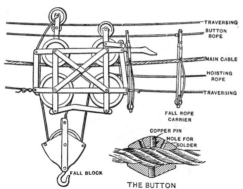

856. FALL ROPE CABLE CARRIER. Miiler type. Five-rope system without fall latch. The power of the hoist rope is increased three times at the fall block.

857. Section of a permanent button stop with pin and holes for soldering.

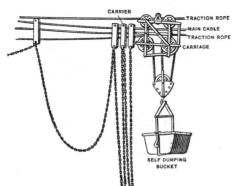

858. FALL ROPE CABLE CARRIER. Four-rope system with loop blocks to support the fall rope.

The traction rope is an endless one, driven from one end of the carrier plant. The lift is three times the power of the fall rope.

859. CRIB DAM. Ottawa River type. A crib framing of timber filled in with stone, topped by a slope frame of 3 to 1, and apron with

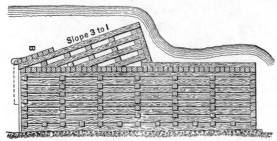

its apex at half the width of the crib to divide the total fall of the water. B, cross planking on top and back. Back filling of stone and earth.

860. COUNTERBALANCED DRAWBRIDGE. Morris Canal type. The draw of the bridge, which is about 25 feet, is manipulated by hand power. The entire length of the bridge is 55 feet. The principle of operation is clearly shown in the illustration. The weight of the draw being about three tons, two counterbalanced weights are employed weighing 3,000 lb. each, made of cast iron in the shape of a cylinder, about 3 feet in diameter, and mounted in such a way as to be rotatable on their axes. These weights run in tracks which are

laid in an ellipse on an inclined framework extending from near the top of the central framework to the level of the roadway of the bridge. Wire cables connect the counterbalance weights with the free end of the draw, the cable passing once around the pulleys at the top of the framework. The pulleys are mounted upon a 3-inch shaft which extends along the top of the central cross beam, and which is provided at its right-hand extremity with a pinion $2\frac{1}{2}$ feet in diameter.   This engages with a small pinion that is mounted on the shaft that is rotated by the endless chain from below.   The draw is almost balanced by the weight of the rollers.

861. TRANSFER BRIDGE. A hanging track on an elevated suspension bridge with a car suspended from a truck and traversed by an overhead cable.

862. EARTH EMBANKMENT.   Ottawa River type.   Solid stone wall on canal side backed by a clay puddle wall and back filling of earth

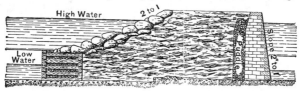

and timber crib work filled with stone and the bank riprapped with large stone.

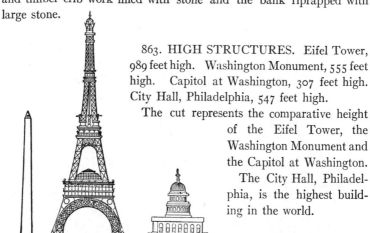

863. HIGH STRUCTURES.   Eifel Tower, 989 feet high.   Washington Monument, 555 feet high.   Capitol at Washington, 307 feet high.   City Hall, Philadelphia, 547 feet high.

The cut represents the comparative height of the Eifel Tower, the Washington Monument and the Capitol at Washington.

The City Hall, Philadelphia, is the highest building in the world.

864. GIGANTIC WHEEL, London, Eng. Three hundred feet in diameter, carried on two towers, 175 feet high, in which are saloon and balconies. The wheel is driven by a steel wire hawser $1\frac{7}{8}$ inches in diameter. There are two of these, one on each side, passing around

grooves on the sides of the wheel, at 195 feet diameter. It is only intended to use one at a time. The motive power is taken from two 50-horse-power dynamos, and of these one will be sufficient, and the other in reserve. There are 40 cars, each 25 feet long, and 8 platforms for loading at once as many cars.

865. End view, showing the three balconies and their lifts.

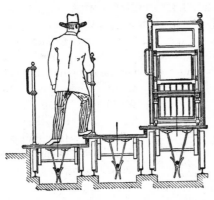

866. MOVING PLATFORM for boarding railway cars. The stepped platform railway will be very safe. Chances of accidents are limited. The fall of a person passing from one platform to another would not be attended with serious results, as the difference between the speed of two platforms is equal to the average speed of a pedestrian.

867. TRAVELING STAIRWAY OR RAMP. A dynamo and a transmission drive the upper drum and guards at a mean speed of twenty

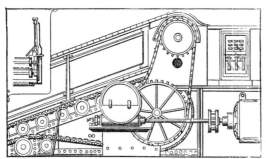

inches per second.

The system com-prises an endless web formed of bars of wood which are provided with rollers that are formed of a material called "hemacite" and that run upon rails. The returning half is suspended from a rail lodged in the lower chord of the principal girder. This arrangement of chains with detach-able links permits of doing away with stretchers.

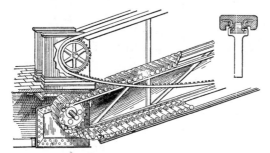

The jointed web is actuated by a chain of which each link cor-responds to one of the bars of wood. This passes at the upper part over an indented wheel actuated by the electric motor with the inter-position of a shaft with a ratchet to prevent any return in an opposite direction.

The jointed bars are provided with rubber projections for the purpose of giving the feet a firm hold. These projections, which are arranged in longitudinal bands, make their exit at the lower part and disappear at the upper between the teeth of metallic combs designed to take up and set down the passengers without jerks. The guards consist also of end-less chains covered with rubber and cloth. Each link of the chain slides in a groove that prevents any lateral displacement.

868. Perspective view, showing jointed web, sprocket drum at the lower end of the ramp and a section of the moving hand rail.

868A. WELL CLEANER. The object of the device shown herewith is to provide means whereby wells and cisterns may be expeditiously cleaned of dirt and sediment without stirring up mud. It comprises a tank adapted to be lowered into the well, and having an ingress pipe in its bottom. An egress pipe projects up from said tank to which a pump is adapted to be connected. Within the tank is a screw for the egress pipe and a valve for closing the ingress pipe. Carried below the tank and surrounding the ingress pipe is an expansible shield of fabric something in the form of an umbrella. This, when the cleaner is on the bottom of the well, is drawn open by chains until it fills the well area. It then acts as a shield to prevent the mud being disturbed by the drawing operation from rising into the water of the upper part of the well.

868B. ELECTRIC BAGGAGE TROLLEY. An electric trolley for the transportation of parcels is shown. The elevated runway consists of a pair of flat steel bars, 12 in. apart, and forms a continuous loop, 2,448 ft. long. The traveling hoist is suspended from the runway by a four-wheeled truck, and the operator rides in a hammock seat suspended just back of the motor. The carrying basket is fitted with small wheels so that it can be moved easily when on the surface.

SECTION XXI.

MISCELLANEOUS DEVICES.

Section XXI.

MISCELLANEOUS DEVICES.

869. PORTABLE SAW for felling trees. The saw is formed of hardened steel plates, which are riveted together in double series for the entire length. The rivets are sufficiently loose to form joints. Each plate or link is shaped on one side to form a pair of saw teeth, one tooth cutting in one direction and one in the other. The plates are a little thicker on the cutting edge than at the back, so that the saw, as it is sharpened, is always set so as to clear its cut. A cross handle at each end of the saw fits into a ring for use. The handles are withdrawn from their rings to render the saw portable.

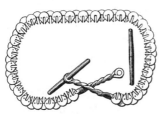

870. STUMP-PULLING MACHINE. The pulling mechanism is supported by a tripod, to the upper end of which is secured a chain carrying a bar or plate provided with a bearing in which slides a notched bar. Meshing with the notches of this bar are the teeth of a pawl, which is so connected by levers with the operating handle that the downward movement of the latter will raise the pawl and notched bar and the chain attached to its lower end. A sliding bolt then holds the notched bar in its raised position, when the handle can be raised to enable the pawl to engage with the next lower teeth of the bar. Thus, by a bar in its raised position, when the handle can be raised to enable the pawl to engage with the next lower teeth of the bar. Thus, by a

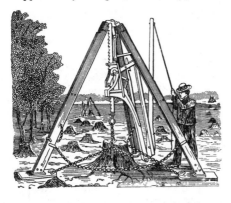

succession of up-and-down movements of the handle, the notched bar may be elevated its entire length, or until the stump is pulled completely out.

871. MOTOR ROLLER-DISK PLOW. The gang of roller

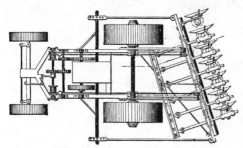

disks are separately attached to arms pivoted to a frame, which is attached to an extension of the rear end of the traction motor. A windlass driven by the motor lifts the disk plows out of the ground when not in use.

872. AUTOMOBILE PLOW. French type. In this system the part designed for working the ground comprises a series of three disks,

which are not arranged in the same plane, although alongside of each other, and each of which carries strong steel colters mounted upon its circumference. These disks are placed upon a frame in the rear of a road locomotive, the mechanism of which is so combined as to set them in rotation. The frame that supports them may be raised more or less, and may also be thrown out of engagement with the earth.

As the locomotive advances, the disks revolve and their peripheral knives penetrate the earth and cut it into slices the thickness of which is capable of varying according to the velocity given the instrument and to the nature of the ground plowed. As may be seen, the colter disks are mounted on the back of the locomotive, and the inclination of the colters as well as the rotation of the disks is in a contrary direction in order to assure the pulverization, which is the real object of plowing. The plow was exhibited at the Paris Exposition, where experiments with it proved that it was capable of plowing six acres per day of twelve hours.

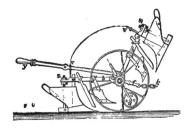

873. REVERSIBLE PLOW. The wheel runs in the last furrow. *y*, yoke handle, which turns over to the other side at the end of a furrow; V, latch to fasten the handle to the beam; *j*, *k*, clevis and chain.

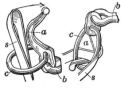

874. TETHERING HOOK. The hook or fastening for tethering or coupling animals by their bridles, etc., and for other uses, comprises a link, *c*, and a hook, *a*, the point of which can not pass through the link. The hook has holes for the link and the fixing-staple *b*. The strap, *s*, is attached to the fastening as shown.

875. FOUNTAIN WASH BOILER. The broad base of the siphon collects the steam generated on the bottom of the boiler, which rising in the vertical pipe induces a rapid flow of boiling water, creating a circulation through the clothes.

876. POTATO-WASHING MACHINE. *a*, spiral of arms for removing dirt; *b*, perforated screw for moving potatoes toward end of

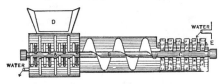

washer next to the comminutor; *c*, perforated paddles for lifting the clean potatoes into the hopper leading to comminutor; *d*, hopper for introducing potatoes into washer; *a*, hopper leading to comminutor, not shown. The machine is slightly tilted so that the water flows to the left, while the potatoes are forced to the right by the screw and spiral arms.

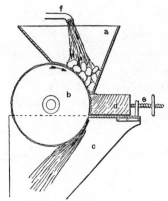

877. POTATO-RASPING MA-CHINE. Used in the manufacture of starch. *a*, hopper ; *b*, barrel rasp ; *c*, receptacle for pulp ; *d*, wooden buffer ; *e*, setting screw ; *f*, water jet.

The buffer is for adjusting the open-ing between the rasping barrel and itself to insure a uniformly fine potato pulp.

878. PARIS-GREEN DUSTER. A small rotary fan with pinion and gear, driven by hand. A vibrating dust box, with a regulating

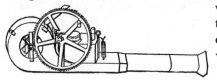

valve and spout. It will dis-tribute a pound of Paris green evenly over an acre of potato vines.

879. AUTOMOBILE MOWING MACHINE. McCormick type. The motor is a double cylinder, 10-horse-power gasoline engine. The

oil tank is divided into three compart-ments : one for oil, one for batteries, and one for water. Power is transmitted from the motor by sprocket wheels and chain to a friction clutch placed on the cross shaft of the mower. This clutch is so ar-ranged as to engage either one bevel-gear wheel or another placed on each side, and in this way the machine can be run backward and forward at will. The bevel-gear wheels engage a pinion which serves to operate the fly-wheel shaft and cutter bar.

They also transmit power to the mower wheels through two gear wheels. The friction clutch is controlled by a lever placed at the foot of the operator. Steering is effected by a crank connected with the guide wheel in front of the cutter bar. The cutter bar can be lifted by the driver from his seat by means of a lever.

880. MODERN TWO-HORSE MOWER. Wood type. All metal construction, except the tongue, whiffletrees, track clearer, and

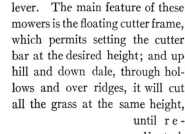

lever. The main feature of these mowers is the floating cutter frame, which permits setting the cutter bar at the desired height; and up hill and down dale, through hollows and over ridges, it will cut all the grass at the same height, until re-adjusted. On all the mowers

the gearing is protected from dust, and roller bearings are used throughout, eliminating all unnecessary friction.

881. CREAM SEPARATOR. Danish type. The milk is fed through the pipe A, and passed down the conical center through tubes

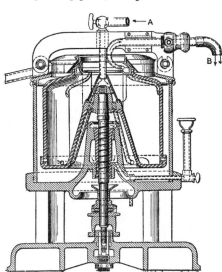

and into the separating pan at the bottom. The cream being of less gravity than the milk separates under the high speed of the pan, and is carried up along the cone and discharged over the top of the revolving pan to the spout at the left hand. The denser milk gathers at the outside of the pan and rises through the openings in the annular groove and is scooped up by the discharge pipe B. About 2,000 revolutions per minute are required in these machines.

882. REFRIGERATION. Ammonia process. The simple routine of the process of refrigeration by the circulation of ammonia. It con-

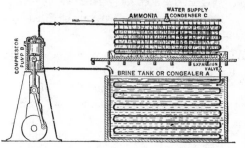

sists of three principal parts: A, an "evaporator," or, as sometimes called, a "congealer," in which the volatile liquid is vaporized. B, a combined suction and compressor pump, which sucks or, properly speaking, "aspirates" the gas or vapor from the evaporator as fast as formed. C, a liquefier or, as commonly called, "condenser," into which the gas is discharged by the compressor pump, and under the combined action of the pump pressure and cold condenser the vapor is here reconverted into a liquid, to be returned to and again used in the congealer.

883. MODEL COLD-STORAGE HOUSE. The lettering on the cut shows the principal features of construction. The ice should

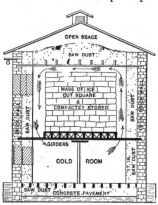

rest on wooden slats laid in a galvanized iron pan a little larger than the pile of ice, with drip pipes and siphons to carry off the water. At the entrance to the storeroom there must be a vestibule, either inside or outside, as space or circumstances may direct. The walls should be thick and the door very heavy. The doors, both inside and outside, should be fitted with rubber, so as to close perfectly tight, and both doors must never be opened at the same time. This vestibule should be large enough to contain a fair wagon load of goods, so that if you are receiving a load of stuff, you are not required to stop until all is in the vestibule and ready to store. This house only needs filling once a year. The temperature will range from 34° in winter to 36° in summer, and will preserve fruit perfectly from season to season. The opening for putting in the ice, shown just under the pulley in the cut, has two doors, with a space between; each door is a foot thick. The window in the cold room has three sets of sashes, well packed or cemented. The walls are 13 inches thick, lined with 17 inches of sawdust. Thirty-six inches of sawdust are put on the floor over the ice. The building shown is 25 feet square, inside measure, and 22 feet from floor of cold room to ceiling over the ice. The ice room is 12 feet high, and the cold room 9 feet. Pillars are required under the center of the ice.

884. MODERN GRAIN HARVESTER. The grain, when cut and

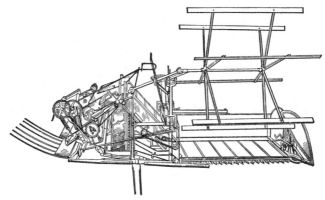

thrown on to the traveling apron, is carried over to the binder, where it is bundled, tied, and dropped on the ground.

885. COMPOUND THRESHER. A threshing-machine study; Reeves type. A, beater drum; B, separator; C, carrier; D, forwarder;

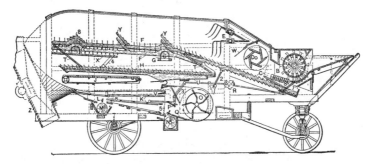

E, push forks; F, push rakes; G, shaking cranks; K, V, winnow sieves; O, winnow fan; M, N, grain chutes.

886. REFUSE CREMATORY. The figure shows a sectional elevation of this destructor furnace in which 1 is the main combustion

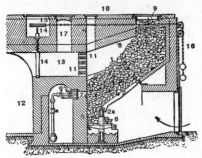

chamber, 2 the fire grate, the lower end of which is carried on a hollow bearer, 2a, through which water is circulated to keep it cool. The lower grate, 6, is of sufficient length to prevent clinker, which falls on to it from the upper grate, from falling over its front end. The clinker remains on this lower grate until more completely burned and partially cooled, when it is raked off over the front end. 7, 7a and 7b are tuyeres through which air, in addition to that entering through the fire bars, is forced; 7 and 7b, on either side of 7a, are not visible in cut; the refuse to be burned is fed through the aperture 9. The stoking hole, 10, permits the introduction of an iron bar for keeping the grate and back wall, 5, from getting clinkered. The products of combustion are drawn through the openings, 11, 11, which are at the hottest part of the fire, into the main flue, 12, through an intermediate chamber, 13, fitted with a damper, 14.

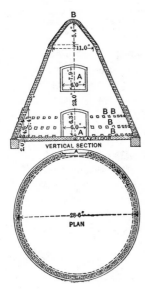

887. CONICAL CHARCOAL KILN. Built round on a clay floor with brick walls 12 inches thick for 7½ feet. Eight-inch wall to top. About 90 vent pipes built into the wall in 3 rows with stoppers. Size of a 35-cord kiln, 28 feet inside at bottom; 28 feet high. A, sheet-iron doors and cast-iron frames, 6x6 feet, or bricked up with mud. Time of burning 9 to 10 days; at 5 days vents are plugged tight. Product of 35 cords, 1,700 bushels. Thirty-five thousand brick are required to construct it.

888. Ground plan of the charcoal kiln.

889. COKING OVEN. Connellsville type. The type now in general use has a diameter of from 10 to 12 feet, and a height of from

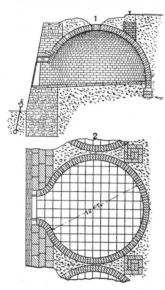

6 to 8 feet, and is built of fire brick or stone. It is arched in the interior, and has an opening in the top for charging and for the escape of the gases during the coking process, and a door in the lower front side through which the finished product is "drawn," this door being closed during the coking process. The average charge of coal per oven is from three and one-half to four tons, the heavier charge requiring more time for the coking process. When the charge is leveled it has a depth of from two and one-half to three feet in the oven, thus leaving sufficient room for the accumulating gas and for the expansion and rising of the coke during the process of its manufacture. It is the practice to charge every other oven each day, and the charge is ignited by the heat retained in the walls of the ovens. The ignition is indicated by a puff something like a powder explosion. For twenty-four hours the gas is allowed to escape, and then the oven is closed up. Furnace coke in general use requires forty-eight hours for the coking process.

890. Plan of one coking oven in a range of a coking plant.

891. DESTRUCTOR FURNACE. English. Vertical and transverse sections of a double line of furnaces. The grates where the fire

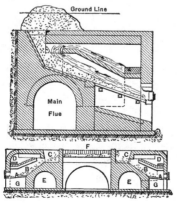

is made are shown at A. At B the refuse to be destroyed is shown in an inclined flue where it is being dried, and as it is consumed on the grate, descends on the slope of the flue, fresh matter being supplied from the pits at C. The down flue, by which the products of combustion are carried to the main flue, E, is shown by the dotted lines in the upper figure.

892. Cross section of a double furnace.

893. LIFE-SAVING NET. For bottom of elevator shafts. A strong rope net, F, held on two sides by the rods, G, which are in turn

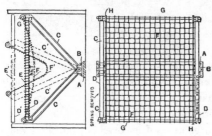

supported on each end by the strut arms C. The lower ends of these arms fit the bearings in the pillow blocks, B, which are bolted to a stout plank, A, which is securely fastened to the bottom of the shaft. The net is held taut by a large compression spring, E, acting at the upper ends of the strut arms C. The spring, E, is supported and held in place by pieces of large pipe, D, it being also free to move along the same.

When the falling body strikes the net, the fall is broken by the combined action of it and the springs which take the position shown by the dotted lines in the vertical section.

894. REMINGTON TYPEWRITER. Vertical transverse section of the No. 3 machine, showing the arrangement of the keys, key

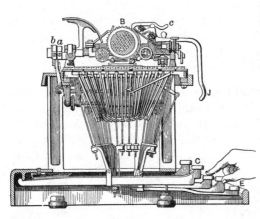

levers, and connections. In the upper part of the main frame of the machine is arranged a ring to which are clamped loops in which are pivoted the type arms. There are in these machines as ordinarily constructed from 38 to 42 type arms, each one bearing at its free extremity a die having on its face two characters, an upper and a lower-case type, figures, and punctuation marks. The type arms are pivoted relative to the ring so that the characters which they bear all strike exactly in the same place. The type arms have hardened steel pivots which are ground to a bearing, thereby insuring accuracy in the movement of the levers.

As shown, each type arm is connected by an adjustable steel wire connector with the key lever pivoted at the back of the machine and projecting beyond the front, where it is curved upwardly and provided with a finger piece or key bearing the character or characters represented by the type arm with which the key lever is connected.

895. REMINGTON TYPEWRITER. The end of the type arm and the double type carried thereby is shown in detail at A, and the paper-

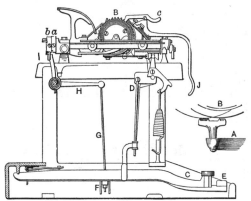

supporting roller, B, is shown in full lines above the lower-case type, and in dotted lines in its position for writing capitals. The capitalizing key, C, which is the foremost one shown in this view, is connected with a right-angled lever, D, through which lateral motion is imparted to the carriage. A spring connected with the lever, D, returns the roller to its normal position as soon as the finger is removed from the capitalizing key. The space bar, E, extends entirely across the front of the keyboard, and a bar, F, which is supported by rods, G, from levers, H, extends under all of the key levers, including the levers attached to the space bar. The levers, H, support the ratchet bar, I, which acts upon the pallets, a, b, in alternation, allowing the spring attached to the paper carriage to move forward one space at a time, as the pallets, a, b, escape from the teeth of the ratchet bar I. When a key is depressed to print a character upon the paper carried by the roller, B, the bar, F, will be moved down and the rack bar, I, shifted from the pallet, b, to the pallet a. This is done without any movement of the carriage ; but when the key is released and the rack bar, I, returns to its position on the pallet, b, it allows the paper carriage to move forward one notch. If a greater space is desired than the normal action of the machine provides, the space bar, E, is touched immediately after printing the character, and if a space is required without writing, the space bar, E, alone is operated.

896. UNITED STATES ARMY AND NAVY GUNS.   Parts of re-enforcement shown in section.   Lengths and sizes shown on cut.

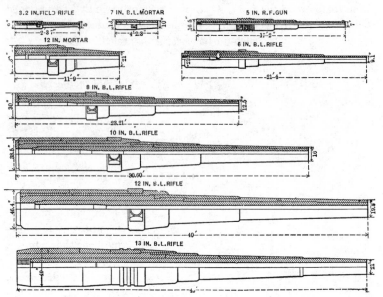

The greatest assumed range of steel rifles of medium sizes is about 12 miles, which requires an elevation of from 40° to 45°, but accuracy of fire is uncertain beyond a range of 4 miles.

897. UNITED STATES MAGAZINE RIFLE.   Krag-Jorgensen type.   The United States magazine rifle is the simplest arm of its kind

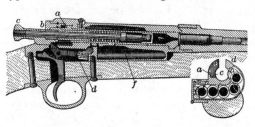

to take apart, as all of the bolt and magazine mechanisms can be dismounted and again assembled without the use of a single tool.

The magazine holds five cartridges, which can be held in reserve by turning the cut-off down; the gun can then be used as a single loader, just as if it had no magazine and, at any moment, the cartridges in the magazine can be fired with wonderful rapidity.

To load this arm, the bolt handle is raised and pulled to the rear in one continuous motion, which operation withdraws the empty cartridge case from the chamber and ejects it from the gun. The top cartridge in the magazine then rises in front of the bolt, if using magazine fire, or a cartridge is dropped in front of the bolt by hand, if using single loader fire, and the bolt handle pushed forward and turned down. This motion seats the cartridge in the chamber and cocks the piece, which is then ready to fire.

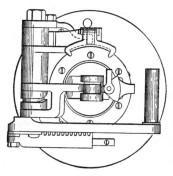

898. BREECH-BLOCK MECH-ANISM for firing large guns. A rack moved by the lever meshing in a sector gear on the breech block revolves the block one sixth of a revolution, when it is swung out of the breech and clear of the bore. The handle near the pivot strikes the extractor lever, which operates the shell extractor and draws the case. Seabury system.

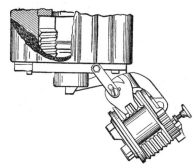

899. Shows the breech block swung clear from the chamber and a section of the screw within the chamber.

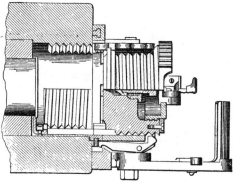

900. The breech block in front of the chamber ready to be pushed in and revolved to a lock position by a further movement of the lever handle.

901. MAGAZINE PISTOL. Luger type. Pressure on a pin at
the butt of the barrel pushes to the rear the barrel and the breech block,

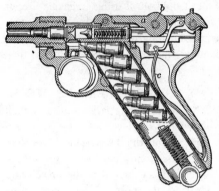

which slides along grooves in
the framework. During this
movement, the movable breech
and the barrel slide as one
piece. The breech, however,
continues to move by its momentum the rollers of the
knee or toggle-joint bearing
against the curved butt piece
of the frame and causing a
circular movement of the link,
a, about its axis *b*. The knee
rises until the moment when the mainspring, *c*, contained in the stock
is entirely compressed, as is also the percussion spring. The cartridge
case carried along by the extractor strikes against the ejector, which
throws it out.

The seat of the breech block being clear, the upper cartridge of the
magazine is pressed by a spring in the magazine in front of the head
cylinder. The mainspring, compressed by the recoil, pushes forward
the breech block through the medium of a stirrup which connects the
two pieces. The knee lowers itself half-way, at the same time communicating its movement to the receiver and to the barrel, while the firing pin
strikes against a lug and the percussion spring remains compressed.

As the knee straightens out, the barrel and the breech block again
act as one piece. The arm is thus again loaded, cocked, and ready to
fire.

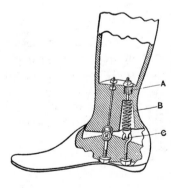

902. ARTIFICIAL ANKLE. The
spring, B, lifts the heel for the forward
movement of the foot ; the pressure of
the body holds the foot in contact with
the ground at the forward movement of
the body. The motion is limited by
the angular space between the solid
bearings.

903. ARTIFICIAL LEG. The socket, A, has a pad, B, and strap, D, for adjusting the size to the stump. K, in 1 and 2, is a bridge piece

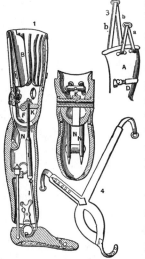

in the upper section resting on the knee bolt, F, and affording the superior point of attachment for the extensor spring, I, and tendons, *i, i,* which throw the foot upward and forward as soon as it is lifted from the ground in walking. The ham strings, N, N, are attached to the posterior portions of the thigh and leg, to act as checks to the forward motion of the leg. The ankle joint consists of a socket in the foot and a ball, P, attached by its neck and the iron frame, Q, to the leg, and has a horizontal stud upon it, fitting its appropriate recess in the socket in the foot, so as to prevent vibration in a horizontal plane, while leaving the joint free for motion in vertical planes, as described. The elastic straps, *a, b* (3), are proportioned as to length and strength, and afford a means of attaching the suspensory yoke (4), whose straps pass over the shoulders, so as not alone to bring the weight upon the framework of the body, but also to enable the shoulders, by their motion, to influence the action of the artificial limb.

903A. ONE-MAN CROSSCUT SAW. The illustration shows how to rig up a large crosscut saw to cut large logs without the aid of another person. The opposite end of the saw is kept from digging too deeply into the wood by a rope run over two pulleys on posts and a weight tied to its end. The weight should be in proportion to the weight of the saw end.

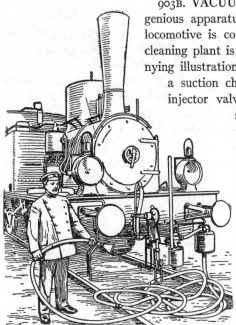

903B. **VACUUM CLEANER.** An ingenious apparatus by means of which a locomotive is converted into a vacuum-cleaning plant is shown in the accompanying illustration. It comprises simply a suction chamber attached to the injector valve of the locomotive, a steam trap for the condensation of water, a dust collector, which is kept partially filled with water, and the necessary amount of hose and types of suction tips or nozzles. The steam trap and dust collector are connected by means of a hose of large diameter, and the suction hose is attached to the bottom of the dust collector.

When the injector valve is opened the live steam passes the opening of the suction chamber and creates a vacuum therein, which produces the suction, drawing the dust and dirt into the dust collector where it is retained by the water. The air, after unloading its burden of dirt, then passes into the steam trap and is exhausted into the atmosphere.

903C. STREET SWEEPER. In this apparatus the entire system of brushes is enclosed within a casing somewhat resembling a carpet

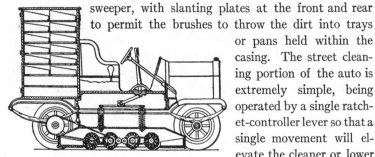

sweeper, with slanting plates at the front and rear to permit the brushes to throw the dirt into trays or pans held within the casing. The street cleaning portion of the auto is extremely simple, being operated by a single ratchet-controller lever so that a single movement will elevate the cleaner or lower it for use.

903C. WRAPPING MACHINE.   A machine for wrapping rectangular packages is shown.   In its general features it is one of that class of wrapping machines wherein the article to be wrapped and the wrapper is inserted in a pocket disposed upon the periphery of a carrying wheel which as it moves brings the article and its wrapper under the action of various instrumentalities whereby paper or other wrapping material is folded down on the article.

The carrying wheel has an intermittent movement equal to the distance between the centers of two of the pockets.   The article to be wrapped is packed within a pocket, a sheet of paper being so placed in

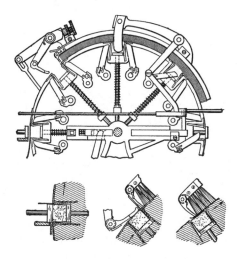

with it as to wrap it on three sides, the ends of the paper projecting out from the pocket.   The rotation of the carrier then brings the pocket beneath a folding finger which as the wheel stops is moved to bend down one of the projecting sides of the paper wrapper forward upon the article, thus forming the first side fold.   If the article has been imperfectly wrapped and projects from the pockets, a spring allows the folder finger to lift and thus prevent breakage.

As the wheel rotates the finger remains in the position shown in the small cut until the operation of bending down the remaining radially projecting portions of the wrapper commences.   This is performed by another finger operating on the same principle as the first one.   An arc-shaped brush is arranged parallel to the rim of the wheel so that the tips of the bristles will exert pressure upon the folded paper and hold it in position until the pocket has so moved that the tucker blades may form the first and second end folds of the wrapper.   The wheel is then moved forward until at the proper time the wrapped article is ejected by plungers.

903D. ROTARY ENGINE, HOFFMAN TYPE.   The cylinder A revolves around the stationary ellipse E, which is permanently attached to the hollow shaft S.   Steam is admitted through the hollow shaft S during the first sixth of the revolution of the cylinder, and passes through the port F into the chamber L where it expands.   The only surface of this chamber which is not rigid is the convex face of the curved seg-

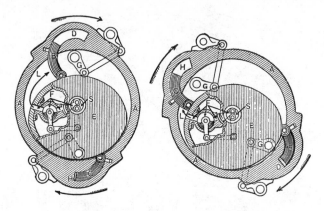

mental blade B, which runs the length of the cylinder, can retreat into its housing D and is fastened by a crank G to the cylinder.   A curved segmental blade is like a section of a pipe cut lengthwise.   Steam pressure on the blade causes it to recede to the right, away from the port F, and thus the cylinder A, which is attached to the segmental blade, is forced to revolve.   As the cylinder revolves it presses down upon the stationary ellipse E, and the blade B is forced back into its housing D just as the blade C is in its housing H at the beginning of the movement. Meanwhile the blade C, which is the duplicate of the blade B, has been carried around and is beginning to protrude as it passes the port F. Steam is again admitted by the automatic cut-off and the same processes are repeated.

SECTION XXII.

DRAUGHTING DEVICES.

Section XXII.

DRAUGHTING DEVICES.

904. GEOMETRICAL PEN. Suardi's type. In the gear proportion, as shown, the diameter of *a* is half of that of A; these wheels are

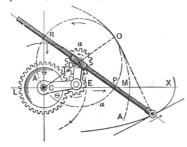

connected by the idler, E, which merely reverses the direction without affecting the velocity of *a's* rotation. The working train arm is jointed so as to pivot about the axis of E, and may be clamped at any angle within its range, thus changing the length of the virtual train arm C, D. The bar being fixed to *a*, then, moves as though carried by the wheel, a^1, rolling within A^1; the radius of a^1 being C, D, and that of A^1 twice as great.

The *ellipse,* then, is described by these arrangements because it is a special form of the epitrochoid; and various other epitrochoids may be traced with Suardi's pen by substituting other wheels, with different numbers of teeth, for *a*.

A number of simple devices for describing elliptic, parabolic, hyperbolic, conchoidal, heliographic and circular curves of great radius, are illustrated and described under the head of "draughting devices" in volume one of Mechanical Movements.

To the professional draughtsman these instruments are valuable adjuncts for delineating, in an easy and satisfactory manner, the delicate and precise curves needed in accurate draughting.

To the amateur, a simple method of projecting geometric curves with precision is a pleasure that stimulates to greater effort in the draughtsman's art.

905. ELLIPSOGRAPH. Mundo type. Will draw ellipses of the smallest size required and of any form from a straight line to a circle.

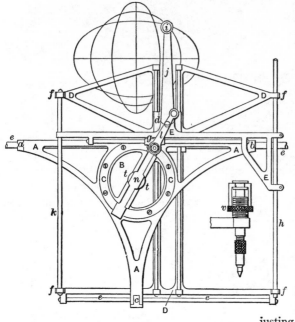

A, main frame with 3 feet at *a*, *b*, *c*. B, crank carrier revolving in the grooved circle C. *l*, the crank. The circular rim, B, carries two slides, above and below, which are clamped at *t* by the thumb-screw. The slides have pivot studs, one of which carries the frame E, and the other the lower frame D; so that by adjusting the two slides and their pivotal connections to the traveling frames E and D, at any distance from the center, *n*, equivalent to the semi-diameters of a required ellipse, the pen, *i*, on the arm, *j*, of the frame, E, will describe the ellipse.

906. Section of the slide carrier and slides, *o*, *p*, and nut, *t*, for clamping them.

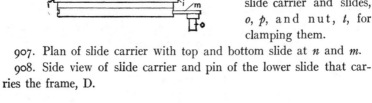

907. Plan of slide carrier with top and bottom slide at *n* and *m*.

908. Side view of slide carrier and pin of the lower slide that carries the frame, D.

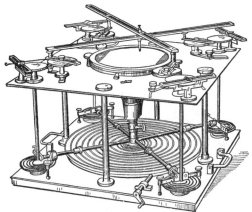

909. THE CAM-PYLOGRAPH. A machine for tracing complex geometrical curves. The small crank on the bottom platform rotates the plate containing a multiple series of gears, which mesh with pinions on four radial arms and transfer their motion through four small but similar gear plates to vertical spindles and to reversing gears on the upper platform.

The face of the gears on the upper platform have trammel pivots to carry the slotted bars that hold the tracing pencil. The tracing table also turns in unison with the gear plate below. The number of loops in the figures are governed by the particular ring gear used.

910. A combination of curves much used in bank-note engraving.

911. Another form of rosette work.

912. Figures formed by a single line tracing.

913. Figure formed **by** four separate line tracings.

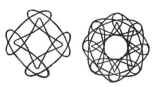

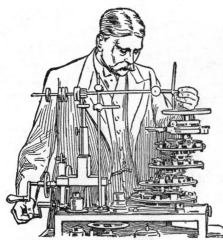

913A. A MACHINE FOR PRODUCING COMPLEX DESIGNS.   A machine by the aid of which any number of intricate pen designs may be produced is shown in the accompanying illustration.   It is called a geometric chuck, and consists of a series of rows or planes of wheels, arranged in tiers. Every wheel, cog and screw is so true in workmanship and arrangement that there is no noticeable friction when the machine is in motion.

SECTION XXIII.

PERPETUAL MOTION.

Section XXIII.

PERPETUAL MOTION.

INTRODUCTION.

THE history of the search for perpetual motion does not afford a single instance of ascertained success ; all that wears any appearance of probability remains secret, and like other secrets, can not be defended in any satisfactory way against the opinions of the skeptical, who have in their favor, in this instance, an appeal to learned authorities against the principle of all such machines, and the total want of operativeness in all known practical results. Published statements afford sorry examples of talents and ingenuity strangely misapplied. Some, but very few, are slightly redeemed from contempt by a glimpse of novelty. Of genius all are deficient, and the reproductions of known fallacies show a remarkable ignorance of first principles on one side and of the most ordinary sources of information on the other. One of the grossest fallacies of the mind is that of taking for granted that ideas of mechanical constructions, apparently the result of accident, must of necessity be quite original. The history of all invention fairly leads to the conclusion that, were all that is known to be swept from the face of the earth, the whole would be reinvented in coming ages. The most doubtful "originality" is that which any inventor attributes to his ignorance of all previous plans, coupled with an isolated position in life. It may be granted that the desire of *secrecy* often renders investigation difficult, and, from some remarkable feeling of this nature, most inventors of supposed perpetual-motion machines, believing themselves possessors of this notable power, make it a matter of profound secrecy.

The attempts to solve this problem would seem, so far, only to have proved it to be thoroughly paradoxical. The inventions resulting from it during the last three centuries baffle any attempt at classification developing progressive improvement. It would almost seem as if each inventor had acted independently of his predecessors ; and, therefore,

frequently reinventing, as new, some exploded fallacy.   These retro-
grade operations and strange resuscitations have led to unmitigated
censure, and a sweeping charge of ignorance, imbecility, and folly. No
doubt many instances exist especially deserving the severest treatment ;
but unsparing censure loses half its causticity, and it shows a weak cause,
or weaker advocacy, to condemn all parties alike as deficient both in
learning and common sense.   It has long been, and so remains to this
day, an unsettled question, whether perpetual motion is, or is not, possible.   To name no other, it is evident, from their writings, that Bishop
Wilkins, Gravesande, Bernoulli, Leupold, Nicholson, and many eminent
mathematicians, have favored the belief in the possibility of perpetual
motion, although admitting difficulties in the way of its discovery.
Against it, we find De la Hire, Parent, Papin, Desaguliers, and the great
majority of scientific men of all classes and countries.   It is evident,
therefore, that even mathematicians are not agreed.

914. PERPETUAL MOTION.   The inventors' paradox.   A demonstration by Dr. Desaguliers in 1719, in regard to the balance of

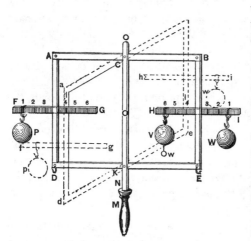

weights at unequal dis-
tances from the center of
oscillation, showing that
the weight P balances the
weight W at any position
on the cross arm H, I, on
the vertical arm B, E,
when pivoted to the
double-scale beam A, B,
and D, E, in which the
resolution of forces is
made apparent in a prac-
tical form so often over-
looked by the inventors
of perpetual-motion ma-
chines.

The cut representing Desaguliers' balance, with his explanation, goes
to show how persistently inventors have ignored the geometrical bearing
of this problem for nearly two centuries.

Desaguliers' Demonstration.—A, C, B, E, K, D is a balance in the form of a parallelogram passing through a slit in the upright piece, N, O, standing on the pedestal, M, so as to be movable upon the center pins C and K. To the upright pieces, A, D and B, E, of this balance, are fixed at right angles the horizontal pieces F, G and H, I. That the equal weights, P, W, must keep each other in equilibrium is evident ; but it does not at first appear so plainly, that if W be removed to V, being suspended at 6, yet it shall still keep P in equilibrium, though the experiment shows it. Nay, if W be successively moved to any of the points, 1, 2, 3, E, 4, 5, or 6, the equilibrium will be continued ; or if, W hanging at any of those points, P be successively moved to D, or any of the points of suspension on the crosspiece, F, G, P will at any of those places make an equilibrium with W. Now, when the weights are at P and V, if the least weight that is capable to overcome the friction at the points of suspension C and K be added to V, as *w*, the weight V will overpower, and that as much at V as if it was at W.

As the lines A, C and K, D, C, B and K, E, always continue of the same length in any position of the machine, the pieces A, D and B, E will always continue parallel to one another and perpendicular to the horizon. However, the whole machine turns upon the points C and K, as appears by bringing the balance to any other position, as *a, b, e, d ;* and, therefore, as the weights applied to any part of the pieces F, G and H, I can only bring down the pieces A, D and B, E perpendicularly, in the same manner as if they were applied to the hooks D and E, or to X and Y, the centers of gravity of A, D and B, E, the force of the weights (if their quantity of matter is equal) will be equal, because their velocities will be their perpendicular ascent or descent, which will always be as the equal lines 4 *l* and 4 L, whatever part of the pieces F, G and H, I the weights are applied to. But if to the weight at V be added the little weight, *w*, those two weights will overpower, because in this case the momentum is made up of the sum of V and *w* multiplied by the common velocity 4 L.

Hence it follows, that it is not the distance, C 6, multiplied into the weight, V, which makes its momentum, but its perpendicular velocity, L 4, multiplied into its mass.

This is still further evident by taking out the pin at K ; for then the weight, P, will overbalance the other weight at V, because then their perpendicular ascent and descent will not be equal.

This "paradox" is illustrated in No. 10, first volume of Mechanical Movements, inviting inquiry by students, a model of which has been exhibited to many doubting amateurs by the author.

915. PERPETUAL MOTION. The prevailing type. A wheel that is furnished at equal distances around its circumference with levers,

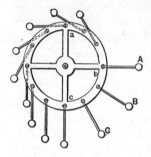

each of which carries a weight at its extremity, and is movable upon a pin, so that in one direction it can lie upon the circumference, while at the opposite side, being carried along by its weight, it may be forced to take the direction of a prolonged radius. This granted, it will be seen that when the wheel revolves in the direction *a*, *b*, *c*, the weights, A, B, C, will deviate from the center, and, acting with more force, will carry along the wheel on this side. And since, in measure as it revolves, a new lever will turn up, it follows, it was said, that the wheel will continue to revolve in the same direction.

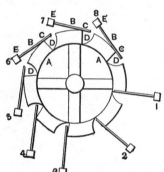

916. PERPETUAL MOTION. Marquis of Worcester. The weights on the ends of the pinioned arms are thrown out as the wheel revolves, giving a greater preponderance by their greater distance from the center of rotation. The precursor of hundreds of motors on the same principle that do not mote.

917. PERPETUAL MOTION. An oft-repeated type, since the times of the Marquis of Worcester. This type has been made with many sections, each section advancing a step in order to overcome its propensity to find a balance and an excuse for stopping.

918. PERPETUAL MOTION. Folding-arm type. The lever, A, is represented in the act of falling from the periphery of the wheel into

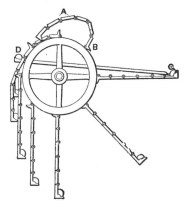

a right line. The lever is composed of a series of flat rods, connected by ruler joints, which said ruler joints are provided with a stop or joggle, to prevent their collapsing at any time more than will bring any one of the rods which compose the levers at a right angle with the rod next to it. This lever is attached to the periphery of the wheel by the hinge joint, B, provided with the shoulder, to prevent its falling into any other than a right line from the center of the circumference of the wheel. The levers are furnished at their outer extremities with a bucket or receiver, the bottom of which is sufficiently broad to retain the ball C. The balls remain in the buckets till the buckets come into the position of the lever, D, when they are expected to roll out of the buckets on to the inclined plane, and by their own gravity roll to the other end of the inclined plane, ready to be again taken into the buckets. Patented in 1821.

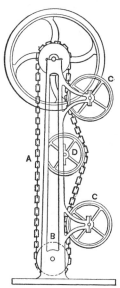

919. PERPETUAL MOTION. Chain wheel. A chain running over the wheels, B, B, is deflected by the idle wheel, D, causing a longer length and weight of chain on that side in proportion to the chain on the straight side A, and like the thousand and one others was expected to go.

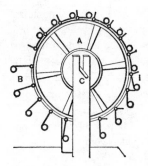

920. **PERPETUAL MOTION.** The most common recurrence of the perpetual-motion idea since the thirteenth century. Inviting to look at, but the resolution of forces in the individual arms and balls demonstrates the equilibrium of forces and its inability to move.

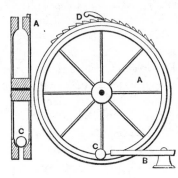

921. **PERPETUAL MOTION.** Magnetism and gravity. B, a strong magnet set in the open slot between the sides of the wheel A, as shown in the section. C, an iron ball. The magnet is supposed to draw the ball to one side of the center, and gravity gives the ball the force to turn the wheel. Patented in 1823.

922. Section showing the ball and slot.

923. PERPETUAL MOTION. The pick-up-ball type. Between the upright frame, A, A, run the wheel, C, geared to the pinion, D,

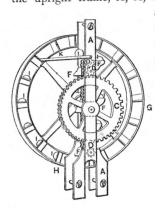

and on the same shaft the two double pinions, D, D, over which double pinions run a double chain, to which chain are fixed the buckets, F, F. The chain is made with joints on each side and bars running across, equal in number to the cogs of the wheel C. Upon the same axle with the wheel, C, on the farther side of the inner stile, A, runs the wheel, G, whose diameter is double that of the wheel C. The wheel, G, is divided near the periphery into receptacles in number equal to the buckets on the chain, which receptacles are supplied with metal balls, I, I, from the buckets, F, F, by means of the gutter, K, which balls by their weight forcing round the wheel, G, and thereby lifting up the buckets, F, F, on one side as they go down on the other side, discharge themselves again at the bucket, L, where they are taken up by the buckets, F, F, and discharged again at the gutter, K, and are so repeated in a constant succession as often as any receptacle is vacant in the wheel, G, at the gutter, K, for their reception, and by that means the perpetual revolution is obtained, the upper ball being at the same time discharged from one bucket when the lower ball is taken up by another.

924. PERPETUAL MOTION. The ball-carrying belt. A represents a wheel with twelve hollow spokes, in each of which there is a

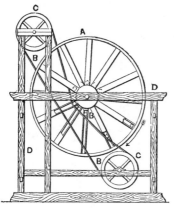

rolling weight or ball. B is a belt passing over two pulleys C. There is an opening round the wheel from the nave to the circumference, so as to allow the belt to pass freely and to meet the weights. The weights are met by the belt as the wheel revolves, and are raised from the circumference until they are at last brought close to the nave, where they remain till, by the revolution of the wheel, they are allowed to roll out through the spokes to the circumference

925. PERPETUAL MOTION. Ferguson's type to prove its impossibility. 1770. The axle is placed horizontally, and the spokes

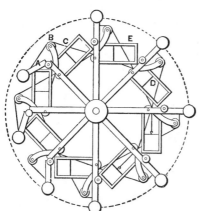

turn in a vertical position. The spokes are jointed, as shown, and to each of them is fixed a frame in which a weight, D, moves. When any spoke is in a horizontal position, the weight, D, in it falls down, and pulls the weighted arm, A, of the then vertical spoke straight out, by means of a cord, C, going over the pulley, B, to the weight D. But when the spokes come about to the left hand, their weights fall back and cease pulling, so that the spokes then bend at their joints and the balls at their ends come nearer the center on the left side.

926. PERPETUAL MOTION. French, 1858. The invention consists in communicating a rotary motion to a fly wheel or drum by

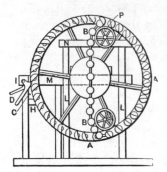

means of a set of falling weights tied together by chains, ropes, or straps. This set of weights, forming an endless chain, runs over two pulleys, suitably disposed up and down near the fly wheel, which is provided with a set of cups fixed around its periphery, so as to receive the weights as they are delivered by the upper pulley, and to carry them down to the lower pulley, whence the same weights reascend in a straight direction to the upper pulley. The weights of the endless chain running or falling down in the curvilinear direction of the periphery of the drum are more numerous than those that are raised up in a straight line, because the curvilinear line is longer than the straight one, and the difference of heaviness due to the number of weights is the force which, by its action at the end of the levers or radii of the drum, causes that drum to rotate.

927. PERPETUAL MOTION. Revolving tubes and balls. The balls, A and B, are in equilibrium because they are at an equal distance

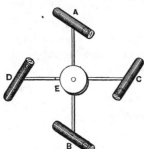

from the vertical line that passes through the center E. By the construction of the machine, the ball, D, being, on the contrary, more distant from the point of support than the ball, C, must prevail over the latter and break the equilibrium. It must then descend to the point, B, and cause the apparatus to make a quarter revolution. Now the latter can not take place without the rod, A, B, which was situated vertically, assuming a horizontal position, and then the balls, A and B, are to each other as were the balls D, C. One must overcome the other and cause the apparatus to make another quarter revolution. This second quarter revolution can not take place without being followed by a third, through the new position assumed by the balls A, B. *Specious argument of the inventor.*

928. PERPETUAL MOTION. Geared motive power. *a* is the axis or shaft on which the wheels are all mounted ; each wheel consists

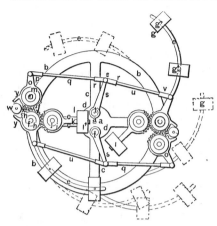

of two parallel rims, *b, b*, each of which is connected by radial arms, *c*, to a boss, *d*, keyed on the axis, *a;* the working parts of each wheel are mounted between the rims and arms thereof, but the outer rim, boss, and radial arms are removed in the figure in order that the working may be fully shown. It must be understood that the pivots or axis, *f, j, n, t*, hereinafter referred to, on which certain parts are mounted, are supported by and extend between the two parallel rims, radial arms, and bosses of the wheel, *b, c, d.   e, e* are curved arms working on axes or pivots, *f*, fixed in the rims; each arm carries weights, *g, g*, held in place by adjusting screws *g'*. Each arm, *e*, terminates at its innermost end in a wheel, *h*, toothed on a portion of its periphery, through which the weight, *v*, forces the weights, *g*, outward at the right-hand side of the wheel, causing a preponderance of weight on that side.

929. PERPETUAL MOTION. The differential hydrostatic wheel. A, B, C, D are four vessels connected to the wheels, E, by round

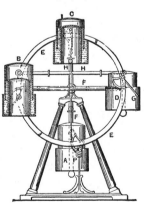

pins which project from the vessels on each side, and enter into corresponding holes in the wheels E. The wheels, E, are intended to revolve by the space under the vessel, B, being a vacuum, and therefore lighter than the same portion of air; a little before the vessel, B, reaches the highest point of the wheels, it begins to close, and opens the opposite vessel, D, in the same manner as the vessel, C, opens A, because the pressure of the atmosphere on the vessel, C, is equal to the pressure on A. Instead of common packing to make the vessels air-tight, mercury is substituted, which has less friction, and is never out of order. The particles of mercury not being entirely free from friction, a little power is requisite to open and shut the vessels; this is expected to be effected by the rods, F, connected to the lever, G, by chains. The rods, F, give motion to other rods, H, by the rollers acting against collars on the rods, H, not shown.

The levers, G, are successively worked by sliding over the roller P. The connecting rods, H, are so adjusted as not to draw the vessels out of their upright position, which would let the mercury escape; also, the lower vessels, A and D, are made rather larger in diameter than B, C, so that the pressure of the atmosphere may counterpoise the weight of the vessels, A, C and B, D, with their connecting rods.

930. **PERPETUAL MOTION**. The lever type. The central weights, A, each weigh one-fourth more than the weights, B, at the ex-

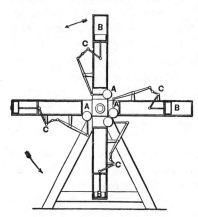

tremities of the arms. The two sets of weights are connected pairs, each pair being joined by a lever, link, and bell crank C. The action of gravity in the central weights compels the sliding weights at the ends of the arms to assume the positions shown in the engraving.

Had this inventor applied a little mathematical calculation to the verification of the truth or falsity of the principle of his device, he might easily have proved that it was a perfect balance, and saved himself both trouble and expense. The leverage of the outside is exactly counteracted by the leverage of the inside weights.

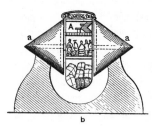

931. **PERPETUAL MOTION**. The fact that a double cone weight will roll uphill on a diverging pair of ways has been taken by a perpetual motionist as the basis for a self-moving car, as shown in the cut, the rails being divergent up grade and parallel down grade in sections. Patented in 1829.

932. PERPETUAL MOTION. The rocking beam. A beam, C, pivoted on a center at D, and connected by a pitman, J, to a crank

and fly wheel, contains a long straight tube at the top and two double inclined tubes below. A ball rolls along the upper tube by gravity in synchronism with the revolving wheel and axle, so that its momentum just carries it to the drop valve and incline at F as the crank reaches the upper point of its revolution. The steeper incline of the lower double bend tube returns the ball to the farther end of the tube in time to start in the straight tube for its next run. Patented in 1870.

933. PERPETUAL MOTION. Tilting tray and ball. This invention consists in the arrangement of an annular tilting tray, which

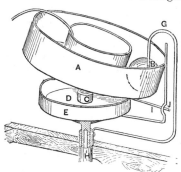

forms the orbit for a revolving ball, in combination with a supporting platform, and with a lever which extends into the tray and connects with a shaft, to which motion is to be imparted in such a manner that, by continually changing the position of the tray, the ball is caused to rotate therein without interruption, and by the action of the rotating ball on the lever the desired motion is imparted to the shaft, which connects with the working mechanism to be driven. A represents a tray, which forms an annular path for the ball B. This tray is made of sheet metal, or any other suitable material, and its diameter is about four times that of the ball B. It is supported in its center by a rod, which connects, by a ball-and-socket joint, C, with a platform, D, so that said tray can be readily tilted in any desired direction. From the edge of the platform, D, rises a circular rim, E, which prevents the tray from being tilted any lower than desirable. U. S. patent, 1868.

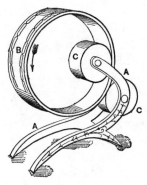

934. PERPETUAL MOTION. The rolling ring which did not roll. It consists of a stand, A, two idler pulleys, C, between which a hollow cylindrical ring, suspended in the manner shown, is expected to revolve in the direction indicated by the arrow. The only difficulty about it is that it will not work, though it looked plausible enough to the inventor.

935. PERPETUAL MOTION. Differential water wheel. From this arrangement it follows that the portion of sponge No. 4 which is

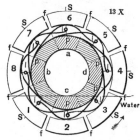

about to quit the water is pressed upon by No. 5 float and spring, which, from acting vertically, is most efficient in squeezing the sponge dry; while that portion of the sponge on the point of entering the water is not compressed at all from its corresponding float No. 8, not having yet reached the edge of the water. By these means, therefore, it will be seen that the sponge always rises in a dry state from the water on the ascending side, while it approaches the water on the descending side in an uncompressed state, and open to the full action of absorption by the capillary attraction.

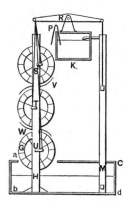

936. PERPETUAL MOTION. Another solution of the water-wheel problem, to be obtained by multiplying the number of wheels, which makes the thing sure to work. The siphon, P, discharges water upon the upper wheel, and by the aprons, V and W, successively to the second and third wheel; all of the wheels are connected to a walking beam by crank and pitman, thus operating a pump for the water supply. Patented in 1831.

937. PERPETUAL MOTION. The gear problem. The frames, B, and the wheels, G, are secured upon the hollow shafts, so that they

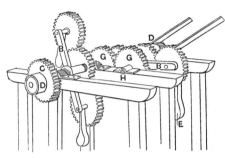

can not move independent of each other. Shafts are placed within the hollow shafts, H, upon which the communicating wheels, D, and the center wheels are secured, so that they can move independent of the frames, B, and wheels, G. While the frames, B, make one revolution, the wheels, D, and the center wheels make two revolutions. This is caused by the action of the weighted levers E. Their weight, or inertia, prevents them from passing around the center of the axis of the wheels with which they are suspended in the revolving frames. The full force of this resistance, or inertia, is applied to the other wheels of each set, and by these wheels communicated to the center wheel.

938. PERPETUAL MOTION. Mercurial wheel. A is the screw turning on its two pivots; B is a cistern to be filled above the level of the

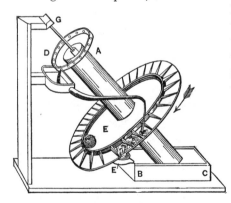

lower aperture of the screw with mercury; D is a reservoir, which, when the screw is turned round, receives the mercury which falls from the top. A pipe conveys the mercury from the reservoir on to the float-board, E, fixed at right angles to the center of the screw, and furnished at its circumference with ridges to intercept the mercury, the momentum and weight of which will cause the float-board and screw to revolve, until, by the proper inclination of the floats, the mercury falls into the receiver, E, from whence it again falls by its spout into the cistern, B, where the constant revolution of the screw takes it up again as before.

939. PERPETUAL MOTION.   Often repeated type. A principle so often employed for the production of self-moving machines that

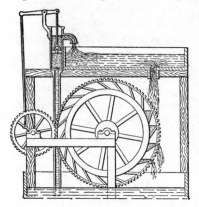

it ranks next to that of perpetual
eccentric weights in its delusive
power upon the minds of inventors.
The attempt to compel a water
wheel to raise the water which
drives it is in one form or other
perpetually recurring in devices
upon which our counsel and opin-
ion are sought. The worst of the
matter is that in most cases our
advice to drop such absurd projects
is received as evidence of want
of sagacity and knowledge, and our
would-be client becomes the dupe of some not over-conscientious patent
agent, who pockets his fees and laughs in his sleeve at the greenness of
the applicant.

The device illustrated is one submitted by one of those enthusiastic
individuals, who, without understanding the first principles of mechanics,
believes he is about to revolutionize the industry of the world by his
grand discovery; and as honor, and not pecuniary reward, is his object.
he seeks to make public his invention through the wide circulation of
some journal.   He is quite willing we should adversely criticise the device, because its merits are so great that no amount of skepticism resulting from our blind prejudice can, he thinks, influence candid minds
against a principle so obviously sound and sublimely simple.

940. PERPETUAL MOTION.   The air-bag problem.   A wheel
with a number of air bags like bellows, fulcrumed on an inner ring and

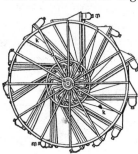

with a weight on the movable cover.   Each
air bag is connected by a tube to the opposite
bag. The wheel is immersed in water,
when the weights compress the air bags at
the left in the cut and extend the bags at
the right side assisted by the hanging weights,
the air passing through the connecting tubes.
Thus, by the inflation of the bags on the right
side, the wheel is made to revolve in the
water.

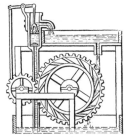

941. PERPETUAL MOTION. A type of one of the many forms of perpetual-motion devices that have been exploited during the past three centuries, and perhaps earlier, in which a water wheel is made to pump the water to drive it.

942. PERPETUAL MOTION. Air transfer in submerged wheel. A, in the cut, is a tank containing water, as shown. The hollow arms,

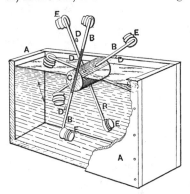

B, communicate with a hollow shaft, C, and the bellows, E, screw valves, D, being employed to increase or diminish the area of the passages in the hollow arms B. Each of the bellows, E, carries a weight, which, during a portion of the revolution, compresses the bellows and forces the air out of it through the hollow arms, B, and shaft, C, into bellows upon the opposite side of the wheel, which, being inverted, are expanded by the action of the weights, and, their buoyancy being thus increased on one side of the wheel, the latter is expected to turn constantly by virtue of the effort of the expanded bellows to rise to the surface.

943. PERPETUAL MOTION. Extending weights and water transfer. The stationary sector gear, A, rolls the small pinions which,

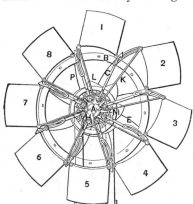

by a rod connection with the following edge of the hinged weights on the periphery of the wheel, tilt the weights upward and outward, making a preponderance on that side of the wheel. The same operation also opens and closes a series of water bags on the inner rim of the wheel, each bag being connected to the opposite bag by a tube, thus adding additional weight to the right-hand side of the wheel.

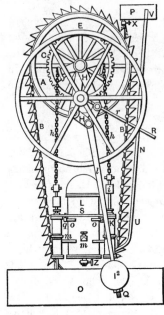

944. PERPETUAL MOTION. English patent (1832) in which a verbose description is given of chain buckets driven by water from a tank, which revolves a geared wheel and pinion and by a cam sustains the vibration of a heavy pendulum, to which is attached a sector beam, pump chains, and counterweights that operate pumps for returning the water to the upper tank.

945. PERPETUAL MOTION. The sponge problem of Sir W. Congreve, of rocket fame. Three horizontal rollers are fixed in a frame;

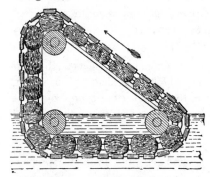

an endless band of sponge runs round these rollers, and carries on the outside an endless chain of weights surrounding the band of sponge and attached to it, so that they must move together, every part of this band and chain being so accurately uniform in weight that the perpendicular side will, in all positions of the band and chain be in equilibrium with the hypotenuse, on the principle of the inclined plane. The frame in which these rollers are fixed is placed in a cistern of water having its lower part immersed.

On the perpendicular side of the triangle, the weights hanging perpendicularly alongside the band of sponge, the band is not compressed by them; and, its pores being left open, the water, at the point where the band meets its surface, will rise to a certain height above its level, and thereby create a load, which load will not exist on the ascending side, because on this side the chain of weights compresses the band at the water's edge, and squeezes out any water that may have previously accumulated in it, so that the band rises in a dry state, the weight of the chain having been so proportioned to the breadth and thickness of the band as to be sufficient to produce this effect.

946. PERPETUAL MOTION. Transfer of air. It is an endless rubber tube, with projections, on which are fastened thin rubber

bags, and a small weight attached to each bag. The bags are filled with air when the weight hangs down, and when it comes on top it presses the air out and through the hollow projection and tube into the next bag that comes in position. When placed over two wheels in water, the bags filled with air should be lighter and rise, while the other side, with the air forced out, should sink.

Each bag, as it comes into position at the bottom of the left tube, will be filled with air expelled from a bag at the top. The weights will descend a certain amount, one in expanding and the other in contracting the bag.

947. PERPETUAL MOTION. Differential weight of balls. The tube is filled one side with water and the other side with enough

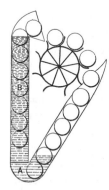

mercury to force the water up to the top of column. In the figure, A is mercury and B the water. The balls to be used are made of iron, with an air-tight chamber filled with gas to make them float in water.

The machine is supposed to operate in this way: The balls are started on the mercury side. Several will be needed to force the first ball through the mercury, but the moment it has passed the center it will rise to the top of column of water. The next coming balls will force it out until it rolls off on to the proper place on the power wheel. Here the balls exert their weight, turn the wheel, and then drop back into the starting channel to force the ones ahead of it through the mercury back into the water again.

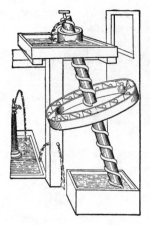

948. PERPETUAL MOTION. Inclined disk and balls. The partitions set at an angle between the outer and inner rim of the wheel roll the balls toward the center on one side and toward the periphery on the other side of the disk. Attached to a screw pump. 1660.

A type of scores of water-raising devices by perpetual motion in the seventeenth and eighteenth centuries.

The Archimedean screw seems to have had a strong hold on the minds of perpetual-motion inventors.

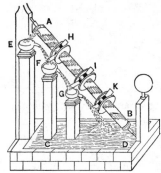

949. PERPETUAL MOTION. Self-moving water power. An Archimedean screw mounted with three water wheels, by its revolution pumps water which falls consecutively upon the wheels, and gives the power required to turn the screw. Seventeenth century.

950. PERPETUAL MOTION. Chain pump as known in 1618. A water wheel which is expected through a system of gearing to operate

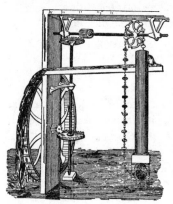

a chain pump, which pump should raise the water necessary to propel the wheel, and so on forever. It is probably unnecessary to inform our readers that this fallacious principle has been tried in various ways, and that there are occasionally yet to be found those so unskilled in mechanical science, and incapable of seeing the radical error of the device, as to waste their substance in a repetition of this time-honored blunder.

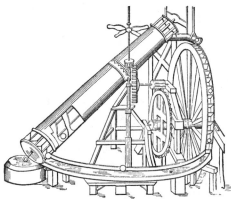

951. PERPETUAL MOTION. The Archimedean screw for raising balls. The balls carried up by the screw were supposed to require less power than they gave by falling on the periphery of the wheel. Enough to drive the screw.

952. PERPETUAL MOTION. Differential weight by flotation. Weights descending through air force themselves by their weight into a

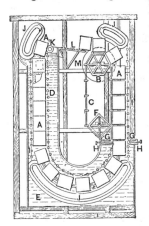

liquid and rise by flotation on the other side of the U-shaped chamber. A represents the blocks; B is the hexagon-shaped wheel; C is the endless chain, which remains attached to the wheel by means of its pointed hooks; E is the receptacle; F is the square wheel from which the chain, C, at the bottom of its course is detached to reascend round the wheel B; G, rollers, of which there are four, made of India rubber or other elastic material, placed at the entrance of the receptacle E; and H is the India-rubber angle pieces, also placed at the entrance, between which rollers, G, and angle pieces, H, pass with slight friction the blocks, after being disengaged from the chain C. These blocks, A, angle pieces, H, and rollers, G, being in close contact, form a stoppage, so that the water can not issue, and are pushed and moved forward by the blocks which descend after them. I is the endless band, resting on supports, J, fixed to the inside of the receptacle, supporting the blocks and moving with them. The blocks, when in the vertical part of the receptacle, are conducted by four wires, one on each of their four sides. K is a roller upon which tilt the blocks, guided by the endless band when on the top of the receptacle to leave the same; L, friction rollers, on which fall and roll the blocks after having tilted, in order to reach the hexagon wheel B.

953. PERPETUAL MOTION. The flotation problem. An upright tank, through which passes a number of floats connected by a

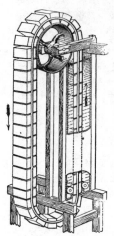

band of elastic rubber attached to their ends, leaving just enough space between them to secure action
on each side by the water. They are each of the
same weight as an equal bulk of water at the surface, therefore the upper one in the tank has no
comparative weight. The next lower one has a
unit of upward force equal to the condensation of
its bulk of water, and so on, each adding a unit to
the upward tendency, until we come to the last, the
pressure on which is altogether downward to the
amount of the entire column of water; but we
already have a number of opposing upward forces,
and when we look on the other side and see the
thirteen active weights, it seems clear that there
will be a large surplus weight, over and above the
opposing weight and the friction of the rollers and upper wheel. The
weights were to pass through an elastic cylinder at the bottom.

954. PERPETUAL MOTION. Liquid transfer. A wheel, each
of whose radii, A, B, contains a small channel through which there is a

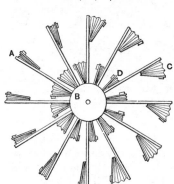

communication between the two bellows, C, D, one of which, C, is at the
extremity of the radius, and the other,
D, is nearer the center. The external
side of these bellows is loaded with a
weight. It will be seen that on one
side (C, for example) the bellows far-
thest from the center must open,
and those nearest must close. A liquid
having been poured into each radius
in sufficient quantity to fill its channel
and one of the bellows, it is evident
that on the side, C, such liquor will be at the extremity, that is to say, in
the bellows that are open, while on the other side it will be in the bellows
that are near the center. Consequently one-half the wheel will be heavier than the other, and so the wheel itself ought to have a perpetual
motion.

955. PERPETUAL MOTION. Chain-pump type. A series of balls placed parallel to each other are hinged or linked together in a

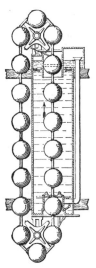

similar manner as the buckets of a chain pump; this chain of floats is passed over two sets of pulleys or disks fixed to two horizontal shafts, the one placed vertically above the other, the said pulleys being formed to suit the diameter of the floats. One-half of this chain of floats passes through the center of the tank holding the water or other fluid, and the other half passes outside the tank through the air. The floats, when in motion, enter through the bottom of the tank, and rise up by their buoyancy through the water; they then pass round the top pulley, descend outside the tank, and, passing over the bottom pulley, again enter the tank, and so on. If cylindrical floats are used, as described, they are fixed on the connecting links half a diameter or more apart from each other. An absurd device is described in this invention of 1865, for opening and closing the entering and exit valves of the chamber and the use of compressed air for operating them.

956. PERPETUAL MOTION. Mercurial displacement in a cistern of water. A cistern full of water 4 feet deep. Let B be a

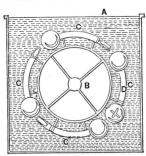

wheel; freely suspended within it, let there be four glass tubes 40 inches long, c, c, c, c, having large bulbs, holding, say, a pint, blown at the closed end. Fill these tubes with mercury, fix on an India-rubber bladder, that will hold a pint, to each of them at the open end, and let them be attached round the wheel, as in the figure. As the pressure of 40 inches of mercury will exceed the atmospheric pressure, and also that of the four-feet column of water, when the India-rubber bottle is lowest, and the tube erect, as at D, the mercury will fill it, leaving a vacuum in the glass bulb above. On the opposite side the mercury will fill the glass bulb, and the India-rubber bottle will be pressed flat, as will also be the case in the two horizontal tubes. Now, it is evident that the two horizontal tubes exactly balance each other; but the tube, D, with its bulb swelled out, displaces a pint of water more than its opposite tube, and hence will attempt to rise with the force of about one pound, and each tube, when it arrives at the same position, must produce the same result; the wheel must have a continual power, equal to about one pound, with a radius of two feet.

957. PERPETUAL MOTION. Air-buoyed wheel. A is a cistern of water filled as high as line R; C are six bladders, communicating 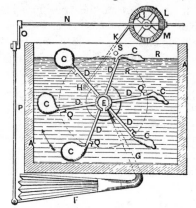 by the tubes, D, with the hollow axle E, which axle is connected with the bellows, F, by the pipe G. H is a crank, connected with the crank, I, by the rod K. L is a bevel wheel, M a pinion, N its shaft. O is a crank attached to the bellows, F, by the rod P. Q are valves with projecting levers. R and S are two projecting knobs. T is a hole in the axle, E, forming a communication with it and the lowermost bladder. The axle, E, being put in motion, is expected to carry round the bladders and tables, and by the cranks, H and I, and the connecting rod, K, cause the wheel, L, to revolve, which, communicating a similar motion to the pinion, M, shaft, N, and crank, O, works the bellows, F, from which the air enters the axle, E, by the tube, G, and passing through the hole in it at T, enters the lower bladder, C, by the tube D; this bladder being thus rendered lighter than the space it occupies, ascends, bringing the bladder behind it over the hole in the axle, T, in like manner, and which is thereby expected to gain an ascending power, producing a similar effect on the one behind it. When one of the bladders arrives at the knob, S, the lever of the valve, Q, strikes against it and opens the valve; when the bladder arrives at C and begins to descend, its pressure on the water drives out the air; the knob, R, then closes the valve, Q, and prevents the entrance of any water into the bladder; by this contrivance, three of the bladders were expected to be alternately full and empty, according as they passed over the hole T or the knob S.

The reason assigned for the failure of this machine was the friction, the old invincible enemy of perpetual-motion seekers.

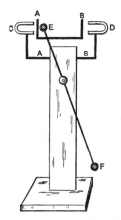

958. PERPETUAL MOTION. By magnetic resistance from alternate interposition of a non-magnetic conductor between the magnets and armature. F, pendulum, E, armature, C, D, magnets. A, B, neutralizing substance moved by the pendulum to a closure between the magnet and armature at the end of the pendulum stroke, alternately, so that the opposite magnet will be in force as the armature swings toward it. Claimed, but not yet found.

959. PERPETUAL MOTION. The overbalanced cylinder. A cylinder containing a fluid with two or more weighted rods passing

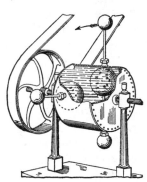

through stuffing boxes in the shell. To the middle of each of these rods is fixed a ball of cork which is expected to rise to the upper side of the cylinder whenever the revolution thereof brings it a little below the axis of the cylinder. In thus rising, it will carry the upper weight away from the center and bring the lower end toward the center so that it is thought the center of gravity of cylinder arms, corks, and metallic balls will be kept constantly on one side of a geometrical center, and constant revolution will result. The fact is, however, that the center of gravity will remain always in a perpendicular drawn through the axis, and, consequently, the expectations of the inventor were never realized.

960. PERPETUAL MOTION. The hydrostatic weight or differential volume problem. A too prevalent belief at the present time

that a large area or body of water has a greater hydrostatic pressure than a connected tube rising from its base. A projector thought that the vessel of his contrivance, represented here, was to solve the renowned problem of the perpetual motion. It was goblet-shaped, lessening gradually toward the bottom until it became a tube, bent upward at *c*, and pointing with an open extremity into the goblet again. He reasoned thus: A pint of water in the goblet, *a*, must more than counterbalance an ounce which the tube, *b*, will contain, and must therefore be constantly pushing the ounce forward into the vessel again at *a*, and keeping up a stream or circulation which will cease only when the water dries up. He was confounded when a trial showed him the same level in *a* and in *b*.

961 PERPETUAL MOTION. Capillary attraction type. Plan and leveation. A tank nearly filled with water and two wheels marked

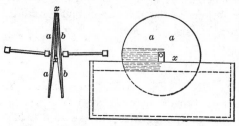

a, a and *b, b* are placed in the water in the tank. By capillary attraction the water rises between the two wheels marked *x, x*, to a height above the level of the water in proportion to the distance of the wheels from each other at *x, x*. As the water rises between the wheels marked *x, x*, above its level, the weight of water between the wheels will cause the wheels to continually revolve.

962. Elevation, showing the position of the water raised by capillary attraction.

963. PERPETUAL MOTION. Magnetic pendulum. Let A, A, represent two magnets revolving on axes. Let B represent a larger

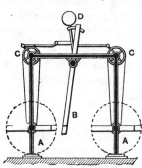

magnet hanging on an axis, pendulum fashion, between the two former. As the poles of the two smaller magnets lie in the same direction, the effect will be to draw the larger magnet toward that on the left hand, while it is at the same time repelled by that on the right; but, while this is going on, the upper end of the large magnet raises, by means of a guide wire, the tumbler D. which, just before the magnets come in contact, passes the perpendicular, and falls over, carrying with it the lever connected with the two wheels, C, C, and causing them to perform a quarter revolution; these wheels are connected by lines with two small wheels fixed on the axles of the two magnets A, A. While the former make a quarter revolution, the latter turn half round. Patented in 1829.

964. **PERPETUAL MOTION.** Magnetic wheel. A light wheel on friction rollers, set with slips of iron at an angle around its periphery.

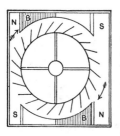

N, N are two magnets, which, attracting the rim of the wheel, will render one side lighter and the other heavier, causing it to revolve *ad infinitum :* or, to render it more powerful, let the steel rims be magnetized, and fixed on the wheel with their north poles toward its center. Let two more magnets be added, as shown by the unshaded lines; let these two, S, S, be placed with their south poles nearest the rim of the wheel, and the other two, N, N, with their north poles in that position. Now, as similar poles repel and opposite poles attract, the wheel will be driven round by attraction and repulsion acting conjointly on four points of its circumference. B, B are blocks of wood to keep off the attraction of the magnets from that part of the wheel which has passed them. No substance yet found that interrupts the magnetic field.

965. **PERPETUAL MOTION.** Magnetic mill of the middle of the eighteenth century. A, B, C, D represents a frame of brass or wood for the machine, E, F, to run in.

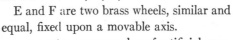

E and F are two brass wheels, similar and equal, fixed upon a movable axis.

1, 2, 3, etc., are a number of artificial magnets placed within the teeth of the wheel all round, and as near each other as is possible, provided they do not touch; their north poles at E and their south poles at F.

H and I are two similar and equal magnets fixed in the brass plate, A, C, very near each other, but not touching.

K and L, two more, fixed in the brass plate, B, D.

Now, as the north pole of one magnet repels the north pole of another magnet and attracts the south, and, inversely, the south pole of one magnet repels the south pole of another and attracts the north, so the south pole, I, attracts all the north ones at E, and the north pole, H, repels all the north ones at M. In like manner, K attracts at N and L repels at O, and by this means the whole machine, E, F, is expected to move perpetually around.

Now this would be all lovely if magnets did not attract in more than one direction.   Many American inventors have tried the same principle over and over, only to find their wheel standing still, and have then sighed for some medium which, interposed between a magnet and its armature, would prevent attraction while thus interposed.

966. PERPETUAL MOTION.   Regenerating pendulum. A, B, E, F is a frame connected by C, D, a crossbar, through which runs g, a pen-

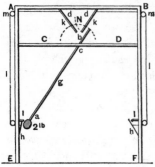

dulum hung on a pivot, C.   This pendulum has two arms, one, a, measuring five feet, and the other, b, one foot in length, connected so together to form a lever with a long and short arm, whose fulcrum is c.   This pendulum has a weight of two pounds at its end.   K, K are two short levers having a joint in them to allow the pendulum to pass them one way, but not the other, without moving them, whose fulcra are d, d, by which they are connected with A, B.   From these run cords, l, l, over pulleys, m, m, which cords are connected (for the purpose of drawing them up into catches) with h, h, springs throwing with a power of three pounds.   I, I are catches for the springs when brought back after working their power. N is the point where the pendulum, g, will escape from the lever K.

967. PERPETUAL MOTION.   Magnetic wheel.   A wheel, A, with a series of armatures on its periphery, revolves before a horseshoe

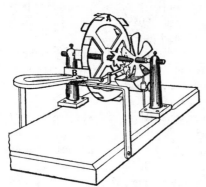

magnet.   Upon the shaft are also mounted a star wheel and a propeller wheel.   The star wheel is arranged to tilt a lever, which carries at its extremity a plate, B, of brass coated with the "chemical and mineral substances" which make it an insulator of magnetism.   The permanent magnet is a U-shaped bar, with its poles near the wheel, A, and opposite the path of the insulating plate B.   The propeller wheel, turning in a cup of water, serves to equalize the motion, and thus prevent the machine from running away with itself and committing self-destruction, so the inventor said.

968. PERPETUAL MOTION. Alternate magnet type. The swinging of the outside magnets of opposite
polarity. Alternate the attraction and
repulsion of the magnets on the wheel
to generate power to swing the outside
magnets in and out of their sphere of
action. Patented 1799.

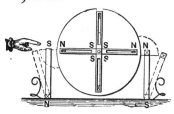

969. PERPETUAL MOTION. Electro-magnetic type. In the
engraving, A represents a frictional electrical machine ; B, a crank ; C,
an electro-magnet ; D, wire con-
ductors ; F, a trunnion ; G, an
armature ; E, a circuit closer ; H, a
pitman ; I, an insulating substance ;
and J, a spiral spring.

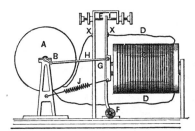

The device is expected to operate
as follows:

The frictional electrical machine
is started, which magnetizes the
temporary magnet and draws the armature toward it. This breaks
the circuit at the point, I, E, which demagnetizes the temporary magnet
and allows the spring, J, to again close the circuit. By this means a
continued motion is expected to be kept up.

To those not familiar with the science of molecular physics this device
may appear very plausible; a little reading, however, upon the subject
of the correlation of forces will serve to show its utter fallacy.

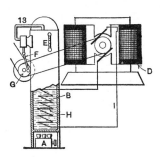

970. PERPETUAL MOTION. Electrical generation. One of the types prevalent among amateur electricians, in which
the electric current from a dynamo is to
generate steam by resistance coils to drive
the engine that runs the dynamo, the
steam being first started by a furnace. F,
engine ; D, dynamo ; B, resistance coil in
boiler H ; A, lamp or furnace.

INDEX.